AF811566

Explosion-Proof Equipment in Hazardous Area

Arvind Kumar Singh

Explosion-Proof Equipment in Hazardous Area

Springer

Arvind Kumar Singh
CSIR-Central Institute of Mining and Fuel
Research
Dhanbad, Jharkhand, India

ISBN 978-981-99-2515-5 ISBN 978-981-99-2516-2 (eBook)
https://doi.org/10.1007/978-981-99-2516-2

© The Editor(s) (if applicable) and The Author(s), under exclusive license to Springer Nature Singapore Pte Ltd. 2023

This work is subject to copyright. All rights are solely and exclusively licensed by the Publisher, whether the whole or part of the material is concerned, specifically the rights of translation, reprinting, reuse of illustrations, recitation, broadcasting, reproduction on microfilms or in any other physical way, and transmission or information storage and retrieval, electronic adaptation, computer software, or by similar or dissimilar methodology now known or hereafter developed.
The use of general descriptive names, registered names, trademarks, service marks, etc. in this publication does not imply, even in the absence of a specific statement, that such names are exempt from the relevant protective laws and regulations and therefore free for general use.
The publisher, the authors, and the editors are safe to assume that the advice and information in this book are believed to be true and accurate at the date of publication. Neither the publisher nor the authors or the editors give a warranty, expressed or implied, with respect to the material contained herein or for any errors or omissions that may have been made. The publisher remains neutral with regard to jurisdictional claims in published maps and institutional affiliations.

This Springer imprint is published by the registered company Springer Nature Singapore Pte Ltd.
The registered company address is: 152 Beach Road, #21-01/04 Gateway East, Singapore 189721, Singapore

Foreword

Explosion-protected equipment, popularly known and symbolized as 'Ex equipment', plays a vital role in the safe operation of coal mines, oil and gas installations, fertilizer and petrochemical plants, defense installations, and other industries where hazardous gases are encountered. The underground coal mines, oil and gas mines, and oil and gas industry are classified as hazardous locations due to the presence of explosive gases, vapors, and combustible dust. The accident and disasters that occur in such an explosive atmosphere indicate that the users of Ex equipment need to be more vigilant in all respect, i.e., design, manufacture, inspection, selection installation, and maintenance.

With rapid advancement in automation and mechanization of coalmines and allied industries, it is necessary from time to time to review and develop new methodology and upgrade testing facilities to keep pace with the latest trend of harmonization of standards for global acceptance of reports and certificates.

The author of this book is heading the Flameproof and Equipment Safety Laboratory of CSIR-Central Institute of Mining and Fuel Research, Dhanbad, since 2003. The book is a compilation of useful information from various sources including national, international standards, international seminars, namely Testing and Certification of Ex equipment (TECEX 2003); Design, Development, Testing and Certification of Ex equipment (DTEX 2009); DTEX 2014; and executive training program. The author of this book has organized all these international seminar and executive training programs in the capacity of Convener and Organising Secretary.

This book covers the overview of hazardous areas containing explosive gases, vapor, and combustible dust, type of explosion protection, mechanism of development explosion pressure, combustible dust, explosion-proof motors, explosions that occur to non-sparking tools, the procedure for testing and certification and its approval, inspection, selection, maintenance, and ignition generated due to coal cutting pick in underground coal mines.

This book has been written with a view to help interested persons to have better knowledge of the standard.

This book will be useful for those who are involved in design, development, specification, installation, commissioning, maintenance, and control of electrical systems.

This is also useful for electrical, mechanical, and mining engineering students, consultants, manufactures, and engineers associated with the explosion-proof equipment and various testing laboratories that test and certify explosion-proof equipment in tackling explosions created by all types of electrical equipment due to arc and spark caused by failure in designing, manufacturing, selection, installation, and maintenance.

B. S. Sonwane
General Manager (HOD)
AME and ISE Division
Bharat Heavy Electricals Limited
Bhopal, India

Preface

As we know, underground coal mines, oil mines, petrochemicals, refinery industries, etc. form the core sector of any country's economy. During extraction of coal from underground coal mines and drilling of oil and consequent its refining explosive gases are frequently encountered, and it creates hazardous area. In such hazardous areas, normal electrical equipment cannot be installed. But a special type of electrical equipment is generally used; these equipment are called Ex equipment or explosion-proof equipment. These explosion-proof equipment play a vital role on breaking the fire triangle with oxygen, flammable substances, and ignition source as its vertices. It is clear that the weakest link in the fire triangle is the ignition source.

The designers, manufacturers, and end users of explosion-proof equipment are in general unfamiliar with the scientific aspects of the principles and functioning of explosion-proof equipment, their design criteria, mode of selection, and maintenance. Now, the Indian Standard has been harmonized as per the International Electrotechnical Commission (IEC). In India, most people in the industry are not familiar with IEC standards. Testing and Certification of Ex equipment as per the IEC code is an essential component of writing this book.

The author of this book, Dr. Singh, has led the Flameproof and Equipment Safety Laboratory of CSIR-Central Institute of Mining and Fuel Research, Dhanbad for a period of almost two decades. Over the years, thousands of tests were conducted under his supervision. At the same time, he organized three international seminars on the subject, wherein he interacted with the foremost leaders and professionals on the planet. He also coordinated more than two dozen executive training programs for the leading petroleum exploration and production company of India, viz., Oil and Natural Corporation Limited (ONGC). This book is a compilation of ideas debated by brilliant minds and useful information from various sources including national and international standards.

The author is grateful to his fellow colleagues and friends like Amit Kumar (for the topic of Intrinsic Safety), Bishwajit Modak, R. K. Mishra, and Md. Kashif Kamal for extending full cooperation in the realization of this dream. This would have never been possible without the moral support of his wife Mrs. Indira Singh and his beloved son Aviraj Singh.

Dhanbad, India Dr. Arvind Kumar Singh

Contents

About the Author

Dr. Arvind Kumar Singh served as Chief scientist at CSIR-Central Institute of Mining and Fuel Research, Dhanbad, where he headed the Flameproof and Equipment Safety Laboratory for more than two decades. He holds a Master's degree in Physics and a doctorate in Nuclear Physics from Banaras Hindu University (BHU), Varanasi. During his academic career, he received an Integrated and Merit Scholarship from the government of Uttar Pradesh. He has published over 110 technical papers in reputed international and national journals, and conferences in the areas of mine ventilation, mine fire, and explosive atmospheres. Currently, he is a member of the ET22 Committee of the Bureau of Indian Standards (BIS). His professional career has included the completion of eight major research projects funded by the Ministry of Coal, Government of India. Over 250 research and consulting projects have been successfully completed under the guidance of Dr. Singh. As a result of these projects, standard products have been successfully transformed into those that comply with Indian standards for hazardous areas. Three international seminars viz., "Testing and Certification of Ex-Equipment in Hazardous Areas" (TECEX-2003), "Design, Development, Testing and Certification of Ex-Equipment" (DTEX-2009 and DTEX-2014) were organized by him as the Organising Secretary. A number of ongoing projects related to Ex 'e', Ex 'p', Ex 'n', explosion-proof cranes, HVAC, and forklifts for hazardous areas are in the pipeline, and the author's expertise and experience will be critical to their successful completion.

Chapter 1
Overview of Hazardous Locations

1.1 Introduction

A hazardous area is one where there is a possibility of fire or explosion due to the presence of flammable liquids, gases, vapors, and combustible dust. The electrical equipment, such as glass fixtures, tube lights, and electric motors, installed in such explosive atmospheres acts as an ignition source. Several national and international codes identify such explosive areas and guide the design of equipment for safe use in such circumstances.

Normally, we do not worry about sparks and arcs generated by electric appliances in our homes, such as tube lights and switches. However, if these appliances are installed in hazardous areas like petrochemical refineries, underground coal mines, and oil mines, they could cause an explosion. This can cause loss of life and destruction of property.

There are many protection techniques by which explosions may be prevented. Our first priority should be to move electrical equipment out of hazardous areas or to properly ventilate hazardous areas with fresh air in order to make them safe.

A special type of electrical equipment is used in hazardous areas such as petrochemical plants, refineries, drilling rigs, pharmaceutical industries, and underground coal mines. This type of equipment is called explosion-proof equipment (Ex equipment). They can be categorized into intrinsically safe, flameproof, and other types of approved categories. Intrinsically safe products prevent explosions by restricting the output energy, while flameproof products quench the flame within the enclosure, i.e., products are not allowed to escape through flameproof joints (i.e., flamepath and gaps). The joint or flamepath acts as a heat sink.

Test houses and codes are available all over the world for testing and certification of explosion-proof products for use in hazardous areas. The American continent follows the National Electric Code (NEC) type of classification, while India and many parts of the world use the International Electro-technical Commission (IEC) type of classification. With the General Agreement on Trade and Tariff (GATT) and Technical Barriers to Trade (TBT) under the World Trade Organization (WTO), it is

© The Author(s), under exclusive license to Springer Nature Singapore Pte Ltd. 2023
A. Kumar Singh, *Explosion-Proof Equipment in Hazardous Area*,
https://doi.org/10.1007/978-981-99-2516-2_1

a requirement that the whole world follows one classification (i.e., IEC type) so that products tested in an approved testing house can be accepted in all countries.

1.2 Technological Approach

In an explosive atmosphere, the primary objective of technology is to reduce the risk of explosion when electrical equipment is used in flammable media. As oxygen and fuel mix together, they form an explosive atmosphere. There is a possibility of an explosion taking place if this explosive atmosphere is exposed to any arc, spark, or hot surface. The first such incident was identified in underground coal mines during the twentieth century. During the year 1971, a paper was presented by Hicks and Brown regarding risk assessment of explosions caused by the ignition of an explosive atmosphere at the conference of electrical engineers [1].

Hence, for any explosion to occur, three things are required, namely flammable substances (such as gas, vapor, mist, or dust), oxygen, and a source of ignition. Therefore, preventing explosions requires breaking the fire triangle in hazardous areas. Oxygen is present in the air in sufficient amounts to support combustion, and it cannot be excluded. The generation of the explosive atmosphere (i.e., a mixture of flammable gas, vapor, mist or dust, and oxygen) is also something that is beyond control, and hence, we need to eliminate the source of ignition. It can be done by properly using explosion-proof equipment (Fig. 1.1).

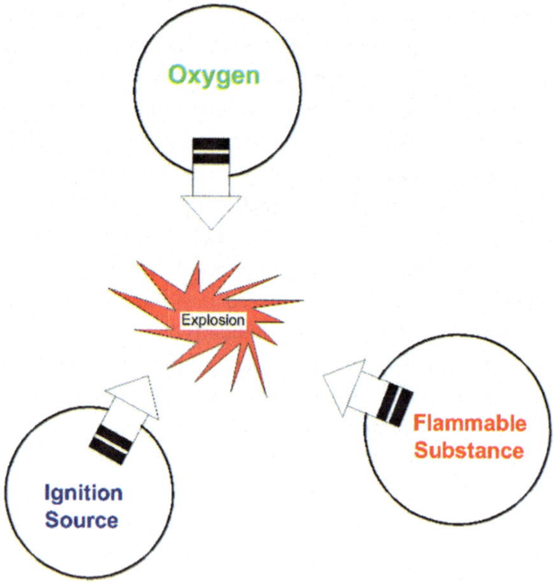

Fig. 1.1 Essential requirements for explosion

1.3 Area Classification

Area classification is used to identify and track explosive atmospheres inside a plant. It reduces the probability of explosion risks and helps in designing the plant. Also, it is used for operational requirements as well as to maintain safety against an explosion in the plant.

However, in some instances, this balance is not maintained because an explosive atmosphere is continuously present in unrestricted areas. People involved in the design and operation of plants need to be aware of the possibility that explosive gas could appear in places where it shouldn't.

1.3.1 Categorization of Sources of Release

The following are the grades of the release of flammable gases [2].

1. **Continuous grade of release**

The point from which flammable gases, vapor, or dust continuously release in the environment and create the explosive zone, i.e., flammable media that persists around 1000 hours per year (more than 10% of the year approx.)

2. **First grade of release**

The point from which flammable gas, vapor, or dust may be released during normal operation into the environment. Thus, an explosive atmosphere is created in this manner. This explosive atmosphere persists between 10 and 1000 h per year (between 0.1% and 10% of the year approx.).

3. **Second grade of release**

The point from which a flammable gas, vapor, mist, and combustible dust may be released for less than 10 hours per year (less than 0.1% of the year).

There are two types of classification prevalent in the world:

- As per the National Electrical Code (NEC) of the USA.
- As per the International Electro-technical Commission (IEC).

1.3.2 Area Classification as Per National Electrical Code of USA (NEC)

In North America, the Hazardous Area Classification system is defined by the National Fire Protection Association (NFPA) 70 [3], National Electrical Code (NEC) 500 to 506 for hazardous areas other than mine, and the Mine Safety and Health Administration (MSHA) for underground coal mines.

1.3.2.1 Classification of Combustible Media as Per the NEC

Class 1: is a location where explosive vapor/gas is present like in petroleum industries.
Class 2: is a location where explosive dust is present like in a grain mill.

Class 3: is a location where loose fibers and flying particles are present, as in textile industries. Mining (methane-firedamp) is not covered under the NEC code but comes under Mine Safety and Health Administration (MSHA).

Class 1 (flammable gases, vapors, and liquids) is categorized into Group A, Group B, Group C, and Group D, and their reference gases are acetylene (C_2H_2), hydrogen (H_2), ethylene (C_2H_4), and propane (C_3H_8), respectively. Group A gases are the most hazardous, and Group D gases are the least hazardous.

Class 2 (combustible dust) is categorized into Group E, Group F, and Group G. Combustible metal dust such as magnesium dust comes under Group E, carbonaceous dust such as carbon and charcoal comes under Group F, and non-conductive dust such as flour, grain flour, a very tiny powder made from wheat comes under Group G.

1.3.2.2 Extent of Hazard as Per the NEC

As per the NEC 70:2008, the extent of hazard is categorized into divisions.

Division 1: Hazardous media can exist under normal operating conditions or due to repair and maintenance or due to leakage caused by faulty operation or breakdown.

Division 2: Hazardous media will exist only under abnormal circumstances due to accidental failures.

1.3.3 Area Classification as Per IEC

The classification of the hazardous area is carried out on the basis of the nature of the release and the condition of the ventilation system in that area. Defining the risk of the presence of flammable media due to gas, vapor, and mists is developed, and such an area is divided into three zones.

In Europe and the rest of the world, except for the American continent, IEC classification prevails. Area classification as per Bureau of Indian Standards (BIS) is also the same as per IEC under the dual marking system. Areas that contain flammable gases and vapor are categorized according to IEC 60079-10-1 [4], and areas that contain combustible dust are categorized according to IEC 60079-10-2 [5]. European countries follow CENELEC standards for area classification.

1.3.3.1 Combustible Media as Per IEC

In accordance with IEC standards, combustible media is classified into two groups, Group I and Group II. Underground coal mines come under Group I, and its reference gas is methane (CH_4) because during the extraction of coal from underground coal mines, methane is emitted. Gases in Group II (such as those produced in the petroleum industry) can be divided into Gr. IIA, Gr. IIB, and Gr. IIC, and its reference gases are propane (C_3H_8), ethylene (C_2H_4), and hydrogen/acetylene (H_2/C_2H_2), respectively. Group IIC gases are the most hazardous, and Group IIA gases are the least hazardous.

1.3.3.2 Extent of Hazard as Per IEC

The petroleum industries like petrochemicals, fertilizers, etc. have flammable gases, vapor, and liquid, and the extent of hazard in these industries is categorized as per IEC 60079-10-1. The classification is useful in the design, construction, and maintenance of electrical equipment in these hazardous areas.

Zone 0: Area in which an explosive atmosphere is present continuously or is likely to be present for longer periods is classified as zone 0 area. Areas like vapor space above the storage process tank and containers both closed and open will fall into this category. IEC 6007911:2011 [6] recommends that in zone 0 hazardous areas, only intrinsically safe products are used. Inherently safe products and their electrical circuits are essentially low voltage and low amperage products and release a small amount of spark energy that is not capable to ignite the explosive atmosphere. This type of device is mounted on the surface rather than the interior of the enclosure.

Zone 1: It is important to note that hazardous media can occur under normal operating conditions, during maintenance, or even during faulty operations or breakdowns. Hazardous areas in zone 1 include the following [7]:

(i) Inadequately ventilated pump rooms for flammable liquids.
(ii) Area around imperfectly fitted peripheral sealing of floating roof tanks.
(iii) Area around flammable liquid or vapor piping systems containing valves, motors, or flange fitting installed inadequately ventilated areas.
(iv) Area likely to contain flammable atmospheric concentration frequently because of maintenance repairs or leakage.

For such locations in addition to flameproof Ex 'd', increased safety motor Ex 'e' or pressurized Ex 'p' motor may also be safe, along with similar non-rotatory equipment of the above category.

Zone 2: An area classified as a zone 2 hazardous area is one in which the occurrence of an explosive mixture is unlikely to occur under normal conditions (vapor is released only under abnormal conditions), or even if it does, it is likely to be temporary in nature. This area is safer than zone 1 and zone 0. Under abnormal conditions, such as when joints or pipelines leak or burst, the area becomes a fire hazard. A standard motor, with additional features, is safe for use in such hazardous areas as non-sparking Ex 'n' and increased safety Ex 'e'.

1.3.4 Parameters Affect the Extent of Hazards

Under Department of Explosive (DOE) regulation, the main considerations which determine the extent of the hazardous area are as follows [7]:

(1) **Type of material**: In determining the extent of hazard, this is one of the most significant criteria. Different types of gases and vapors are ignitable in different composition ranges. Also, depending on the gas, the lower explosive limit (LEL) varies. Consequently, the extent of the hazardous area also varies accordingly. The rate of diffusion is also an important characteristic of the material. Those gases with a higher diffusivity will dilute to compositions below LEL much earlier than others, which will result in a reduction in hazardous zones.
(2) **Vapor density of the material:** The vapor density of the flammable gas determines whether the gas is heavier or lighter. Without walls or enclosures, heavier gases or vapors travel downward and outward, while lighter gases/vapors travel upward and outward.
(3) **Pressure of process or storage:** The extent of hazard changes significantly if flammable gases or vapors are released under pressure. When the pressure at the point of release is high, gases/vapors travel a long distance before they diffuse out to a concentration below the lower explosive limit.
(4) **Size of release:** There are multiple points in an equipment or plant from where flammable vapor or gases are released. This includes the number of flange connections to a process or storage tank or piping. In large plants with large equipment, flow rates are high. Therefore, we should opt for overall plant classification. For small plants, a single source may be an option.
(5) **Ventilation conditions:** The extent of the hazard depends on the ventilation condition of that hazardous area. If the area is open and well-ventilated, then the hazardous area categorized in zone 1 may be converted into zone 2.

As per the fourth schedule of Petroleum Rules 2002, the extent of hazard when plant equipment released under normal conditions and abnormal conditions is shown in Figs. 1.2 and 1.3, respectively.

1.4 Comparison Between IEC and NEC

Following are significant technical differences in defining the zone and divisions [8].

1. The IEC allows the use of only Ex 'ia' category equipment in zone '0', but the NEC does not impose any such restrictions.

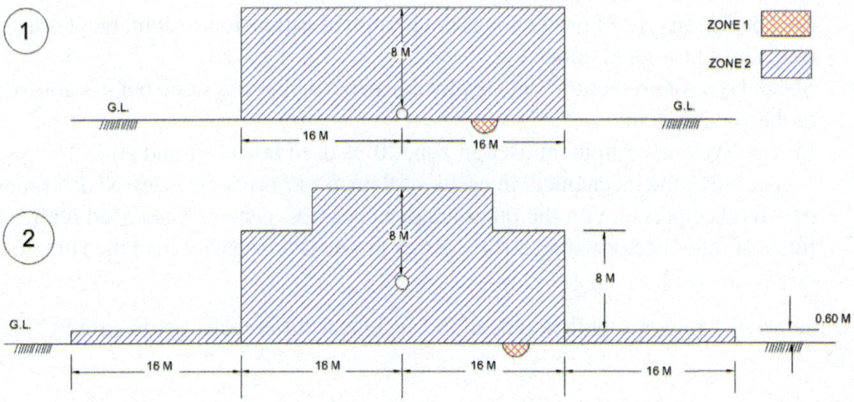

Fig. 1.2 Plant equipment—release under normal conditions—open air. (1) At ground level. (2) Above ground level—large release

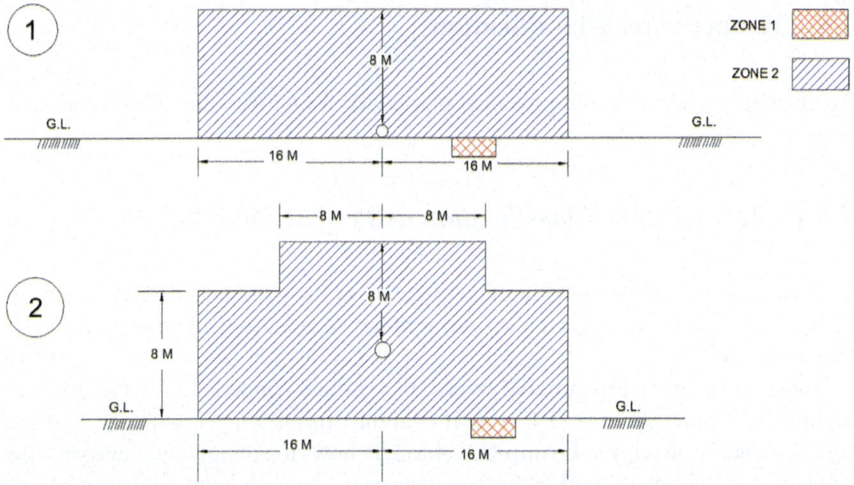

Fig. 1.3 Plant equipment—release under abnormal conditions—open air. (1) Source at ground level. (2) Source above ground level

2. As per NEC, purging and pressurization are carried out in all locations, but as per IEC, it is not possible in zone '0' location.
3. As per IEC, only Ex 'ia' product with two fault conditions is allowed in zone system, and as per NEC, lesser form safety with one fault, Ex 'ib' product is allowed in division system.
4. The division system utilizes the concept of flamepath in Division 1 and Division 2 locations. However, the zone system does not allow this concept to the implemented in zone 0.

5. Encapsulation type of protection is not permitted in division system, but it is used frequently in zone system.
6. Special type of protection is not applicable in the division system, but it is allowed in the zone system.
7. In zone system, equipment used in zone ‘0’ is used in zone 1 and zone 2.
8. As per NEC, the mechanical integrity of flameproof products is tested at 4 times of reference pressure on the other hand, as per IEC system, it is tested with 1.5 times of reference pressure, so the former products are heavier than the latter and costlier as well.

These are the regional philosophical and technical differences between NEC and IEC.

It is important to know the differences before taking any business or investment decision.

1.5 Temperature Classification

There are two classifications prevalent worldwide for temperature classification.

1.5.1 Temperature Classification as Per IEC 60079-0:2017

In flammable media, the maximum temperature of the exposed surface of electrical equipment should not exceed the auto-ignition temperature of flammable gases, vapor, or dust.

Group I: In underground coal mines, the surface temperature of the electrical equipment should not exceed 150 °C if coal dust forms a layer and 450 °C if coal dust does not form a layer. **Group II**: Table 1.1 shows the temperature classification of electrical equipment used in surface industries like petroleum industries. As per IEC, classification temperature is categorized as follows [8].

The temperature classification depends on ambient temperatures ranging from −20 to 40 °C. The classification of a mixture of gases, vapor with air is carried out on the basis of their Maximum Experimental Safe Gap (MESG) and minimum ignition temperature (MIC) and material characteristics for gas and vapor classification as explained in the IEC 60079-12:2012 [9] and IS/IEC 60079-20:2010 [10] codes, respectively. For electrical equipment that is certified by the Certifying Agency for temperature class T4, its surface temperature should not exceed 135 °C. This temperature is calculated based on the ambient temperature of 40 °C. If the equipment is flameproof Ex ‘d’, then the component temperature inside the enclosure may be above the surface temperature, but on classification of temperature class, the surface temperature of electrical equipment is only considered because due to the design of flameproof products, the flame cannot escape through the flamepath and gap.

Table 1.1 Temperature classification as per IEC

Identification number	Maximum surface temp. (°C)	Gases and vapors with respect to temperature classifications
T6	85	Carbon disulfide
T5	100	No vapor or gas specified as yet
T4	135	N-hexane, tetrahydrofuran, n-heptane, N-tetradecane, trichlorosilanece, ethyl glycol, n-nonane acetaldehyde, ethyl ether
T3	200	Petrol, crude oil, n-propyl alcohol, turpentine, acetylene, cyclohexane
T2	300	Ethanol, ethyl acetate, ethane, methane, acetone, toluene, ethylene, cyclohexanone, iso-amylacetate,1, 4-dioxan, n-butane, n-butyl alcohol acetic acid, butaq 1,3, diene, ethyl chloride xylene benzene (pure), vinyl acetate, ethyl benzene
T1	450	Water gas, ammonia, coal gas, hydrogen, carbon monoxide, chlorobenzene

1.5.2 Temperature Classification as Per NEC

Table 1.2 shows the temperature classification as per NEC 500 [11] code for surface industries, i.e., for Classes I and II apparatus.

Table 1.2 Temperature classification as per NEC 500 for Classes I and II apparatus

Identification number	Degrees C (Max.)
T1	450
T2	300
T2A	280
T2B	260
T2C	230
T2D	215
T3	200
T3A	180
T3B	165
T4	135
T4A	120
T5	100
T6	85

1.5.3 Temperature Classification as Per NEC 503 [12] for Class III Apparatus

- The maximum temperature for apparatus not subjected to overloading is 165 °C.
- The maximum temperature for apparatus subject to overloading is 120 °C.

1.5.4 Harmonization of NEC Codes with IEC

The National Electrical Code (NEC) of the USA has a separate code NEC 505 [13] for classification identical to IEC. This code was introduced with a view to allow harmonization of the IEC and NEC codes.

1.6 Product Code

The manufacture, certification, installation, use, and inspection of explosion-proof equipment (Ex equipment) in explosive atmospheres are complex processes. To begin with, it is imperative that we understand the standards (codes) by which this Ex equipment is designed, as well as the code-laying body that formulates these codes. In India, the Bureau of Indian Standards (BIS) is the code-laying body, while the International Electro-technical Commission (IEC) is the international code-laying body. European Union (EU) countries are subject to IEC and CENELEC standards. The CENELEC standards have been aligned with IEC standards. The Underwriters Laboratories Inc. (UL) and Factory Mutual Research (FM) are two independent organizations responsible for framing product codes, testing certifications, conformity assurance, and monitoring products. The Factory Mutual (FM) system follows the same pattern as the Underwriters Laboratories (UL). The Nationally Recognized Testing Laboratory (NRTL) in the USA is an independent organization that tests and certifies Ex-products. The Canadian Standard Association (CSA) formulates product codes, and electrical equipment is also tested and certified in accordance with that code in order to be used in hazardous areas safely.

1.7 Testing and Certification of Products

Many test houses test electrical equipment used in hazardous areas. The Flameproof and Equipment Safety Laboratory of the Central Institute of Mining and Fuel Research (CSIR-CIMFR), Dhanbad, is the first recognized laboratory in India for testing and certification of Ex equipment. CSIR-Central Institute of Mining and Fuel Research (CSIR-CIMFR), Dhanbad, is a NABL-accredited laboratory in India that was established in the early 1960s. As per the standards established by the International Electro-technical Commission (IEC), it is fully equipped with modern equipment received from the UK under the INDO-UK collaboration program for the testing and certification of hazardous equipment used in mining, oil and gas, and allied industries.

Additionally, this laboratory has the capability to test and certify non-electrical products as well, such as flame arresters, PVC and steel cord belts, brattice cloths, non-sparking tools, and frictional incendivity tests for light aluminum alloy for safe use in hazardous environments as these products when used in flammable media also pose fire and explosion hazards.

Other testing laboratories include the Central Power Research Institute (CPRI) in Bangalore, the Electronic Regional Testing Laboratory (ERTL) in Kolkata, Karandikar Laboratories Pvt. Ltd., Mumbai, Intertek India Ltd., New Delhi, UL and FM in the USA, CSA in Canada, PTB in Germany, etc.

1.8 Quality Control of Product

Under the Bureau of Indian Standards (BIS) Act 1986, the manufacturer is granted a license to mark explosion-proof products.

Ex equipment used in Group I, IIA/IIB combustible media and intrinsically safe Ex 'i' are included in the licensing scheme of the Bureau of Indian Standards. The flameproof product Ex 'd' used in IIC combustible media and other explosion-proof products do not fall within the scope of the BIS licensing scheme.

Coal mine clause 157(4) 1957 requires a BIS license for electrical equipment Ex 'd' for underground coal mine use. Under the coal mines regulation 1957, the Directorate General of Mines Safety (DGMS), Dhanbad, enforced that gassy mines must use flameproof equipment. The DGMS grants approval for field trials once the products have been certified by any of the test houses. After the products have been tested and a report has been received from the users, the product is approved. For underground coal mines, the DGMS has the authority to grant approval for explosion-proof equipment.

1.9 Selection of Ex Equipment

In hazardous areas, electrical equipment should be selected and installed in accordance with IEC 60079-14:2013 [14] and inspected and maintained in accordance with IEC 60079-17:2013 [15]. Repairs and overhauls are performed in accordance with IEC 60079-19:2017 [16]. UL in the USA is an independent organization that tests and certifies Ex-products for use in hazardous areas. In the UK, testing and certification of electrical equipment used in hazardous areas are carried out in accordance with British or European Standards. Testing and certification of electrical equipment are carried out by two certification authorities like HSE and SIRA as per their own terms and conditions. The HSE or SIRA markings are fixed on the electrical equipment to indicate the certification details. The ATEX directive became mandatory in countries of the European Union on June 30, 2003, requiring the CE mark for all electrical and non-electrical products that will be used in hazardous environments. If its mark is absent, then the equipment cannot be used in such hazardous areas [17]. If an Indian manufacturer wishes to export their products to the European Union, they should follow the CENELEC standard and obtain the CE mark from the European Agency. In other words, the CE mark is a passport that allows manufacturers to export their products to the EU.

1.10 Equipment Protection Level (EPL)

As shown in Table 1.3, it defines the level of protection for electrical equipment based on the likelihood of it becoming a source of ignition.

Table 1.4 shows the apparatus, group, ignition temperature class, vapor density compared with air, flash point, and flammability limit in air with respect to compound [18].

Since 2004, the Bureau of Indian Standards (BIS) has harmonized the hazardous area standards with the corresponding standards of the International Electro-technical Commission (IEC). Table 1.5 shows the latest IEC standards for different applications in hazardous areas.

1.11 Ex Marking for Explosive Gas Atmospheres

The Ex marking shall include the following:

(a) The symbol Ex which stands for explosion-proof.
(b) The symbol for the level of protection.

Flameproof:

i. 'da' Ex 'd' enclosure (for equipment protection level Ga or Ma).
ii. 'db' Ex 'd' enclosure (for equipment protection level Gb or Mb).
iii. 'dc' Ex 'd' enclosure (for equipment protection level Gc).

Increased Safety:

i. 'eb' Ex 'e' (for equipment protection level Gb or Mb).
ii. 'ec' Ex 'e' (for equipment protection level Gc or Mc).

Intrinsic Safety:

i. 'ia' Ex 'i' (for equipment protection level Ga or Ma).

Table 1.3 Equipment protection level

EPL Ma	EPL Mb	EPL Ga	EPL Gb	EPL Gc	EPL Da	EPL Db	EPL Dc
'Very high' level of protection equipment used in underground coalmines susceptible for firedamp	'High' level of protection used in underground coalmines susceptible for firedamp	'Very high' level of protection equipment used in explosive gas atmosphere	'High' level of protection equipment used in explosive gas atmosphere	'Enhanced' level of Protection equipment used in explosive gas atmosphere	'Very high' level of protection equipment used in combustible dust atmosphere	'High' level of protection equipment used in combustible dust atmosphere	'Enhanced' level of protection equipment used in combustible dust atmosphere

Table 1.4 Apparatus group and temperature class for common flammable gases***

Compound	Apparatus group	Ignition temp. (°C)	T class	Vapor density compared with air (air = 1)	Flash point	Flammability limits in air	
						lower vol. %	upper vol. %
1	2	3	4	5	6	7	8
Methane (firedamp)	I	595	T1	6.07	–	5	15
Acetaldehyde	IIA	140	T4	1.52	– 38	4	57
Acetic acid	IIA	485	T1	2.07	40	5.4	16
Acetone	IIA	535	T1	2.0	– 19	2.15	13
Acetyl acetone	IIA	340	T2	3.5	34	1.7	–
Acetyl chloride	IIA	390	T2	2.7	4	5.0	–
Allyl chloride	IIA	485	T1	2.64	< – 20	3.2	11.2
Ammonia	IIA	630	T1	0.59	–	15	28
Amphetamine	IIA	–	–	4.67	< 100	–	–
Amyl acetate	IIA	375	T2	4.48	25	1.0	7.1
Amyl methyl ketone	IIA	–	–	3.94	(49)	–	–
Aniline	IIA	617	T1	3.22	75	1.2	8.3
Benzene	IIA	560	T1	2.7	– 11	1.2	8
Benzaldehyde	IIA	190	T4	3.66	65	1.4	–
Benzyl chloride	IIA	585	T1	4.36	< 60	1.2	–
Bromobutane	IIA	265	T3	4.72	21	2.5	–
Bromoethane	IIA	510	T1	3.76	< – 20	6.7	11.3
Butane	IIA	365	T2	2.05	– 60	1.5	8.5
Butanol	IIA	340	T2	2.55	29	1.7	9.0
Butyl acetate	IIA	370	T2	4.01	22	1.4	8.0
Butyl amine	IIA	(312)	T2	2.52	– 9	–	–
Butyl digol	IIA	225	T3	5.59	78	–	–
Butyl methyl ketone	IIA	(530)	(T1)	3.46	23	1.2	8
Butyraldehyde	IIA	230	T3	2.48	< – 5	1.4	12.5
Chlordimethyl ether	IIA		–	–	–	–	–
Chlorobenzene	IIA	637	T1	3.88	28	1.3	7.1
Chlorobutane	IIA	(460)	(T1)	3.2	< 0	1.8	10.1
Chloroethane	IIA	510	T1	2.22	–	3.6	15.4
Chloroethanol	IIA	425	T2	2.78	55	5	16
Chloroethylene	IIA	470	T1	2.15	–	3.8	29.3

(continued)

Table 1.4 (continued)

Compound	Apparatus group	Ignition temp. (°C)	T class	Vapor density compared with air (air = 1)	Flash point	Flammability limits in air	
						lower vol. %	upper vol. %
Chloromethane	IIA	625	T1	1.78	–	10.7	13.4
Chloropropane	IIA	520	T1	2.7	< – 20	2.6	11.1
Coaltar neptha	IIA	272	T3	–	–	–	–
Cresol	IIA	555	T1	3.73	81	1.1	–
Cyclobutane	IIA	–	–	1.93	–	1.8	–
Cyclohexane	IIA	259	T3	2.9	– 18	1.2	7.8
Cyclohexanol	IIA	300	T2	3.45	68	1.2	–
Cyclohexanone	IIA	419	T2	3.38	43	1.4	9.4
Cyclohexene	IIA	(310)	(T2)	3.83	< – 20	1.2	–
Cyclohexy lamine	IIA	290	T3	3.42	32	–	–
Decahydronaphthalene	IIA	260	T3	4.76	54	0.7	4.9
Diacetone alcohol	IIA	640	T1	4.0	58	1.8	6.9
Diaminoethane	IIA	385	T2	2.07	34	–	–
Diamyl ether	IIA	170	T4	4.45	(57)	–	–
Dichlorobenzene	IIA	(640)		5.07	66	2.2	9.2
Dichloroethane	IIA	440	T2	3.42	– 10	5.6	16
Dichloroethylene	IIA	(440)		3.55	– 10	9.7	12.8
Dichloropropane	IIA	555	T1	3.9	15	3.4	14.5
Eiethlamine	IIA	(310)	(T2)	2.53	< – 20	1.7	10.1
Diethylaminoethanol	IIA	–	–	4.04	(60)	–	–
Diethyl oxalate	IIA	–	–	5.04	76	–	–
Diethyl sulfate	IIA	–	–	5.31	104	–	–
Dihexyl ether	IIA	185	T4	6.43	75	–	–
Di-isobutylene	IIA	(305)	(T2)	3.87	(2)	–	–
Dimethylamine	IIA	(400)	(T2)	1.55	–	2.8	14.4
Dimethylaniline	IIA	370	T2	4.17	63	1.2	7
Ethane	IIA	515	T1	1.04	–	3.0	15.5
Ethanol	IIA	425	T2	1.59	12	3.3	19
Ethanolamine	IIA	–	–	2.1	85	–	–
Ethyl acetate	IIA	460	T1	3.04	– 4	2.1	11.5

(continued)

Table 1.4 (continued)

Compound	Apparatus group	Ignition temp. (°C)	T class	Vapor density compared with air (air = 1)	Flash point	Flammability limits in air	
						lower vol. %	upper vol. %
Ethylbenzene	IIA	431	T2	3.66	15	1.0	6.7
Ethyldigol	IIA	–	–	4.62	94	–	–
Ethylformate	IIA	440	T2	2.55	– 20	2.7	16.5
Ethyl mercaptan	IIA	295	T3	2.11	< – 20	2.8	18
Ethyl methyl ketone	IIA	505	T1	2.48	– 1	1.8	11.5
Formdimethylamide	IIA	440	T2	2.52	58	2.2	16
Hexane	IIA	233	T3	2.79	– 21	1.2	7.5
Hexanol	IIA	–	–	3.5	63	1.2	–
Heptane	IIA	215	T3	3.46	– 4	1.1	6.7
Kerosene	IIA	210	T3	–	38	0.7	5
Met aldehyde	IIA	–	–	6.07	36	–	–
Methane (industrial)	IIA	–	T1	–	–	–	–
Methanol	IIA	455	T1	1.11	11	6.7	36
Methyl acetate	IIA	475	T1	2.56	– 10	3.1	16
Methyl acetoacetate	IIA	280	T3	4.0	67	–	–
Methylamine	IIA	430	T2	1.07	–	5	20.7
Methyl cyclohexane	IIA	260	T3	3.38	– 4	1.15	6.7
Methyl cyclohexane	IIA	295	T3	3.93	68	–	–
Methyl formate	IIA	450	T1	2.07	< – 30	5	23
Naphtha	IIA	290	T3	2.5	– 6	0.9	6
Naphthalene	IIA	528	T1	4.42	77	0.9	5.9
Nitrobenzene	IIA	480	T1	4.25	88	1.8	–
Nitromethane	IIA	415	T2	2.11	36	–	–
Nonane	IIA	205	T3	4.43	30	0.8	5.6
Nonanol	IIA	–	–	4.97	75	0.8	6.1
Acetaldehyde	IIA	–	–	4.42	52	–	–
Octagon	IIA	–	–	4.5	81	–	–
Paraldehyde	IIA	235	T3	4.56	17	1.3	–
Pentane	IIA	285	T3	2.48	< – 20	1.4	8.0
Pentanol	IIA	300	T2	3.04	34	1.2	10.5

(continued)

Table 1.4 (continued)

Compound	Apparatus group	Ignition temp. (°C)	T class	Vapor density compared with air (air = 1)	Flash point	Flammability limits in air	
						lower vol. %	upper vol. %
Petroleum	IIA	–	–	–	– 20	–	–
Phenol	IIA	605	T1	3.24	75	–	–
Propane	IIA	470	T1	1.56	–	2.0	9.5
Propanol	IIA	405	T2	2.07	15	2.15	13.5
Propyl amine	IIA	(320)	(T2)	2.04	– 20	2.0	10.4
Propylene	IIA	(455)	(T1)	1.5	–	2.0	11.7
Propyl methyl ketone	IIA	505	T1	2.97	(16)	1.5	8.2
Pyridine	IIA	550	T1	2.73	17	1.8	12.2
Styrene	IIA	490	T1	3.6	30	1.1	8.0
Toluene	IIA	435	T1	3.18	6	1.2	7
Toluidine	IIA	480	T1	3.7	85	–	–
Trimethylamine	IIA	–	–	3.5	< 0	1.2	8
Trimethylamine	IIA	(190)	(T4)	2.04	–	2.0	11.6
Trim ethylbenzene	IIA	470	T1	4.15	–	–	–
Turpentine	IIA	254	T3	–	35	0.8	–
Xylene	IIA	464	T1	3.66	30	1.0	6.7
Acrylonitrile	IIB	480	T1	1.83	– 5	3	17
Alleyne	IIB	–	–	1.38	–	1.7	–
Butadiene	IIB	430	T2	1.87	–	2.1	12.5
Butane	IIB	440	T2	1.94	–	1.6	10
Carbon monoxide	IIB	605	T1	0.97	–	12.5	74.2
Cyclopropane	IIB	495	T1	1.45	–	2.4	10.4
Deputy ether	IIB	185	T4	4.48	25	1.5	7.6
Diethyl ether	IIB	170	T4	2.55	< – 20	1.7	36
Dimethyl ether	IIB	–	–	1.59	–	3.7	27.0
Diplopy ether	IIB	–	–	3.53	< 21	–	–
Dioxide	IIB	379	T2	3.03	11	1.9	22.5
Diesoline	IIB	–	–	2.55	(2)	–	–
Epoxy propane	IIB	430	T2	2.0	< 20	2.8	37
Ethoxyethanol	IIB	235	T3	3.1	95	1.8	15.7
Ethyl acrylate	IIB	–	–	3.45	9	1.8	–
Ethylene	IIB	425	T2	0.97	–	2.7	34

(continued)

Table 1.4 (continued)

Compound	Apparatus group	Ignition temp. (°C)	T class	Vapor density compared with air (air = 1)	Flash point	Flammability limits in air	
						lower vol. %	upper vol. %
Ethylene oxide	IIB	440	T2	1.52	–	3.7	100
Ethyl methyl ether	IIB	190	T4	2.07	–	2.0	10.1
Formaldehyde	IIB	424	T2	1.03	–	7	73
Hydrogen sulfide	IIB	270	T3	1.19	–	4.0	75.6
Isopropyl nitrate	IIB	175	T4	–	20	2	100
Methoxymethane	IIB	285	T3	2.63	39	2.5	14
Methyl acrylate	IIB	–	–	3.0	– 3	2.8	25
Nitromethane	IIB	410	T2	2.58	27	–	–
Nitropropane	IIB	420	T2	3.06	(49)	–	–
Paraformaldehyde	IIB	300	T2	–	70	–	–
Tetrahydrofuran	IIB	(260)	(T3)	2.49	– 17	2.0	11.8
Tetra hydrofurfuryl alcohol	IIB	280	T3	3.52	70	1.5	9.7
Town gas (coal gas)	IIB	–	T1	–	–	–	–
Trioxide	IIB	410	T2	3.11	(45)	3.6	29
Blue water gas	IIC	–	T1	–	–	–	–
Hydrogen	IIC	560	T1	0.07	–	4.0	75.6

ii. 'ib' Ex 'i' (for equipment protection level Gb or Mb).
iii. 'ic' Ex 'i' (for equipment protection level Gc).

Encapsulation:

i. 'ma' Ex 'm' (for equipment protection level Ga or Ma).
ii. 'mb' Ex 'm' (for equipment protection level Gb or Mb).
iii. 'mc' Ex 'm' (for equipment protection level Gc).

Non-sparking:

i. 'nA' Ex 'n' (for equipment protection level Gc).
ii. 'nC' protected sparking (for equipment protection level Gc).
iii. 'nR' restricted breathing (for equipment protection level Gc).

Oil filled:

i. 'ob' Ex 'o' (for equipment protection level Gc).
ii. 'oc' Ex 'o' (for equipment protection level Gc).
iii. 'op is' optical radiation, inherently safe (for equipment protection level Ga or Gb or Gc or Ma or Mb).
iv. 'op pr' optical radiation, protected fiber (for equipment protection level Gb or Gc or Mb).

Table 1.5 IEC codes for hazardous areas

IEC code no.	Title of IS/IEC standard
IEC 60079-0:2017	Explosive atmosphere Part 0: Equipments-General Requirements
IEC 60079-1:2014	Explosive atmospheres Part 1—Equipment Protection by Flameproof Enclosure 'd'
IEC 60079-1-1 2002	Electrical Apparatus for explosive gas atmosphere Part 1-1: Flameproof enclosures 'd'—Method of test for ascertainment of maximum experimental safe gap
IEC 60079-2:2014	Explosive atmosphere Part 2: Equipment Protection by Pressurised enclosure 'p'
IEC 60079-13:2017	Electrical Apparatus for explosive gas atmospheres Part 13: Equipment protection by pressurized room 'p' and artificially ventilated room 'v'
IEC 60079-16:1990	Electrical Apparatus for explosive gas atmospheres Part 16: Artificial Ventilation for the protection of analyser houses
IEC 60079-5:2015	Explosive Atmospheres Part 5: Equipment Protection by Powder filling 'q'
IEC 60079-6:2015	Explosive Atmospheres Part 6: Equipment Protection by Oil immersion 'o'
IEC 60079-7:2015	Electrical apparatus for explosive gas atmospheres Part 7: Increased Safety 'e'
IEC 60079-25:2020	Electrical Apparatus for explosive gas atmospheres Part 25: Intrinsically Safe Systems
IEC 60079-26:2021	Explosive atmospheres Part 26: Equipment with Equipment Protection Level (EPL) Ga
IEC 60079-27:2008	Explosive atmosphere Part 27: Field bus intrinsically safe concept (FISCO)
IEC 60079-28:2015	Explosive atmosphere Part 28: Protection of Equipment and transmission system sing optical radiation
IEC 60079-15:2017	Electrical Apparatus for explosive gas atmospheres Part 15: Construction Test and Marking Type of protection 'n' Electrical Apparatus
IEC 60079-18:2014	Electrical Apparatus for explosive gas atmospheres Part 18: Construction Test and Marking Type of protection 'm' Electrical Apparatus
IEC 60079-10-1:2020	Explosive gas atmospheres Part 10-1: Classification of area
IEC 60079-10-2:2015	Combustible Dust Atmosphere Part 10-2: Classification of area
IEC 80079-20-1	Material characteristics for gas and vapour classification—test methods and data
IEC 60079-14:2013	Electrical atmospheres (Other than mines and explosives) Part 14: Electrical installations Design, Selection and Erection
IEC60079-17:2013	Explosive Atmosphere (other than mines and explosives) Part 17: Electrical Installation Inspection and Maintenance
IEC60079-19:2019	Explosive Atmosphere (other than mines and explosives) Part 19: Equipment Repair, overhaul and reclamation

(continued)

Table 1.5 (continued)

IEC code no.	Title of IS/IEC standard
IEC 60079-35-1:2011	Caplights for use in mines susceptible to firedamp—General requirements—Construction and testing in relation to the risk of explosion
IEC 60079-35-2:2011	Caplights for use in mines susceptible to firedamp—Performance and other safety-related matters
IEC60079-29-1:2016	Explosive atmosphere Part 29-1: Gas Detectors-Performance requirements of detectors for flammable gases
IEC60079-29-2:2015	Explosive atmosphere Part 29-2: Gas Detectors-Performance requirements of detectors for flammable gases and oxygen
IEC60079-30-1:2015	Explosive atmosphere Part 30-1: Electrical resistance trace heating—General and testing requirements
IEC60079-30-2:2015	Explosive atmosphere Part 30-2: Electrical resistance trace heating—application guide for design, installation and maintenance
IEC60079-31:2022	Explosive atmosphere Part 31: Equipment dust ignition protection by enclosure 't'
IEC 80079-20-2:2016	Explosive atmospheres—Part 20-2: Material characteristics—Combustible dusts test methods

v. 'op sh' optical radiation, interlock (for equipment protection level Ga or Gb or Gc or Ma or Mb).

Pressurized:

i. 'pv' Ex 'p' (for equipment protection level Gb or Gc).
ii. 'pxb' Ex 'p' (for equipment protection level Gb or Mb).
iii. 'pyb' Ex 'p' (for equipment protection level Gb).
iv. 'pzc' Ex 'p' (for equipment protection level Gc).

Sand filled:

i. 'q' Ex 'q' (for equipment protection level Gb or Mb).

Special protection:

i. 'sa' Ex 's' (for equipment protection level Ga or Ma).
ii. 'sb' Ex 's' (for equipment protection level Gb or Mb).
iii. 'sc' Ex 's' (for equipment protection level Gc).

1.11.1 Certificate and Marking Information

The certificate number and marking information are significant for end users as safety is dependent on it. But there is no shortcut method to the identification of suitable equipment. Putting the marking and label on equipment is the responsibility of the manufacturer of the equipment. The following are the standard coding practices that appear on the certificate and label.

Certifying body CSIR-CIMFR, Dhanbad
Certificate number: CMF 22 INEx 0219
CMF = Central Institute
22 = Certifying Year
INEx = International Ex equipment
0219 = No. of issued certificate in the current year

Certification Marking: This is an important information about the electrical equipment that shows the information of standard, type of protection, gas group, and temperature class. This appears as follows:

EEx d IIC T_4

These symbols stand for:

E—indicates that equipment comply with CENELEC standard
Ex—indicates this product is explosion protected
d—indicates the type of protection Ex 'd', i.e., flameproof
IIC—indicates that flameproof product tested IIC gas group (i.e., mixture of hydrogen with air and acetylene with air) atmosphere
IIA—equipment is tested in mixture of propane and air
IIB—equipment is tested in mixture of ethylene and air
i—intrinsic safety
o—oil filled
p—purge protected
q—sand filled
d—flameproof enclosure
e—increased safety
m—encapsulation
n—non-sparking
T4—indicates the maximum surface temperature of equipment which should not exceed the auto-ignition temp. of the flammable media. Temperature class is generally measured at an ambient temperature of 40 °C.

References

1. Hicks WF, Brown KJ (1971) Assessment of explosion probability for intrinsically safe apparatus. In: TEE conference on electrical safety in hazardous environments 16–18 March, 1971
2. McMillan A (1998) Area classification. In: Electrical installations in hazardous areas (first published 1998)
3. NFPA 70:2018, Standard for electrical safety in the workplace

4. IEC 60079-10-1:2020 Explosive atmospheres—part 10-1: classification of areas—explosive gas atmosphere
5. IEC 60079-10-2:2015 Explosive atmospheres—part 10-2: classification of areas—explosive dust atmosphere
6. IEC 60079-11:2011 Explosive atmosphere—part 11-equipment protection by intrinsic safety "i"
7. Kaul RC (2003) Area classification & electrical equipment selection under department of explosive regulations—an overview. In: An international seminar on "testing and certification of Ex Equipment in hazardous area—a global approach" TECEX-2003. Organised at CSIR-CIMFR, Dhanbad
8. Chakraborty T (2014) Global testing and certification requirements for the 3-phase medium voltage induction motors suitable for hazardous area. In: An international seminar on "design, development, testing and certification of ex-equipment. Organised at Kolkata in India, DTEX-2014, pp 183–198
9. IEC 60079-12:2010—Electrical apparatus for explosive gas atmosphere—classification of mixtures of gases or vapours with air
10. IEC 60079-20: 2010—Material characteristics for gas and vapour classification-test methods and data
11. NEC 500: Temperature class of explosive atmosphere in the form of gas/vapour/dust)
12. NEC 503: Temperature class of explosive atmosphere in the form of ignitable fibre
13. NEC 505: Hazardous area classification
14. IEC 60079–14:2013—Explosive atmospheres—part 14: electrical installations design, selection and erection
15. IEC 60079-17:2013—Explosive atmospheres—part 17: electrical installations inspection and maintenance
16. IEC 60079–19:2019—Explosive atmospheres—part—19: equipment repair, overhaul and reclamation
17. Achari J (2003) ATEX directive and its implication in India. In: An international seminal "Testing and certification of ex equipment in hazardous area—a global approach" TECEX-2003. Organised at CSIR-CIMFR, Dhanbad
18. Sarkar SK (1988) General introduction to electrical equipment in explosive atmosphere. In: Flameproof equipment design, construction, certification use and maintenance

Chapter 2
Type of Explosion Protection

2.1 Introduction

As we read in the first chapter, if we want to prevent an explosion in a hazardous area, we should focus on breaking the fire triangle with oxygen, flammable substance, and ignition source as its vertices. Oxygen is present in abundance in our atmosphere, and there is no dearth of flammable substances like methane and coal dust clouds in the underground mines. To substantiate, during the extraction of coal from underground coal mines, methane is emitted and it is explosive in nature between 5 and 15% v/v with air and has been responsible for a large number of explosions in the country in the last five decades [1]. Similarly, during the drilling of oil and subsequent refining and storage, explosive gases are encountered. Exploiting fossil fuel efficiently mandates the use of electrical equipment in these hazardous areas and any arcing, sparking, or hot surface provides the source of ignition and completes the fire triangle. Thus, the use of electrical equipment in the coal mines and petroleum industries cannot be safe unless electrical equipment is not explosion-proof.

It is clear from the above discussion that the weakest link in the fire triangle is the ignition source. Therefore, if we want to prevent an explosion, eliminating the ignition source should be our focus. There are eight types of protection by which the generation of the source of ignition can be prevented. These are namely,

1. Flameproof
2. Intrinsically safe
3. Purge protected
4. Increased safety
5. Non-sparking
6. Encapsulation
7. Oil-filled
8. Sand-filled

In this chapter, detailed discussions will be carried out related to the performance requirements, construction features, applications, and testing methodology of these

© The Author(s), under exclusive license to Springer Nature Singapore Pte Ltd. 2023
A. Kumar Singh, *Explosion-Proof Equipment in Hazardous Area*,
https://doi.org/10.1007/978-981-99-2516-2_2

types of protected products. The concept of electrical equipment protection consists of four main techniques [2].

- Allow the explosion to occur inside the enclosure but prevent its spread into the outer atmosphere.
- Explosion may be prevented by eliminating arc, spark, and hot surfaces so that the source of ignition is not present.
- Explosion can be prevented by restricting the output energy so that the spark energy cannot ignite the explosive atmosphere.
- The explosion may be prevented by keeping flammable gases out of enclosures.

2.2 Flameproof or Explosion-Proof, Ex 'd'

Flameproof protection is one of the oldest protection techniques which is suitable for use in zone 1 and zone 2 hazardous areas. A series of pioneer research was carried out to develop effective and economical designs for flameproof equipment. The first time, in 1902, Mr. Davis developed a flame-tight DC motor, but it was not air tight. In 1908, Dr. Ing Carl Beyling gave the concept of flameproof enclosure to the mining industry. For this pioneering work, Dr. Ing Carl Beyling was honored by the Institution of Mining Engineers U.K. in 1938 [3].

The term FLP originated from the mining industry for the first time it was used. As per IEC 60079-1:2014 [4], a flameproof enclosure may be defined as 'an enclosure for electrical apparatus that will withstand an internal explosion of flammable gas or vapor without suffering damage and without communicating combustible product and flame in the outer atmosphere'. Thus, a flameproof product is one that allows an explosion to happen inside the apparatus but prevents its spread into the outer atmosphere.

2.2.1 Design Parameters of Flameproof Equipment

Gas group: There are four gas groups, i.e., Gr. I for underground coal mines and Gr. IIA, IIB, and IIC for surface industries like petrochemical, fertilizer, refineries, drilling rig, etc.

Equipment certified for gas Gr IIC are considered suitable for use in combustible media like Gr I, Gr IIA/IIB, based on the permission granted by the Directorate General of Mine Safety (DGMS) for gas Gr I and Petroleum Explosives and Safety Organization (PESO) for gas Gr IIA/IIB (as applicable in India).

Flamepath and Gap

The length between two meeting surfaces (e.g., body and cover) in a flameproof enclosure is known as Flamepath. The space left between these two surfaces after assembly is called a gap. It works as a heat sink for the flame. The unit of measurement of flamepath and gap is in mm.

Threaded length: The threaded flameproof joint is generally used for cable entries, and enclosures with plain covers; a minimum of 8 mm is required for enclosures larger than 100 cc. The minimum requirement of threads engaged is 5, and the pitch size may be between 0.7 and 2 mm for parallel threads and 0.9 mm pitch size for tapered threads with a minimum 6 number of full threads.

Materials of construction: For underground coal mines, the materials of construction should be cast iron, or SS or MS light metal alloy products are not applicable for belowground coal mines.

Mechanical Strength

Strength of enclosure: The flameproof enclosure which envelopes the electrical circuit of components should be able to sustain explosion pressure without transmission of flame through joints. It should be hydraulically checked for 1.5 times the pressure of the reference explosion pressure, for a duration of a minimum of 10 s. No. Swelling, damage, deformation, or leakage should be observed. It should be checked for impact energy as specified in IEC60079-0:2017 [5]. Fasteners are used to maintain the mechanical integrity of the enclosure and maintain the gap as per the specifications. Fasteners or bolts are provided with plain or spring washers to prevent loosening in service. The minimum yield stress required for these fasteners is 240 N/mm^2.

Inspection windows and light-transmitting parts: Glass is used for the inspection of internal components like a voltmeter, ammeter, etc. Similarly, round or dome-shaped glass is used for floodlights or well-glass lighting fittings. A cementing compound like epoxy resin should be used for fixing glass from the inside only. A retaining ring or plate should be attached to hold the glass in place. Glass cement should not lose its properties in the entire range of service temperatures. The glass part should be heat and impact resistant and of toughened quality. A wire guard with a maximum mesh size of 50×50 mm^2 can be provided. The minimum impact energy for use in coal mines is 7 J for glass parts provided without a guard.

Cable entries: The electrical wires and cables connected to the apparatus are fitted with double compression flameproof cable glands. A cable gland may be tested and certified flameproof separately. It should sustain cable pull-out and torque tests. The cable gland must also undergo an impact test after being attached to the enclosure. A cable gland comprises a threaded nipple for attachment with flameproof enclosure, a body consisting sealing (neoprene) ring for compression between the cable sheath and itself, a cone and ring for clamping armor wires of the cable, and an outer cap with a rubber bush for weatherproof protection against ingress of water and dust into the enclosure.

A schematic diagram of a flameproof cable gland is shown in Fig. 2.1.

Terminal box: Cables or wires are terminated separately in a separate enclosure known as a terminal box. Additionally, flameproof motors have a cable sealing box to seal the cable before it is terminated in the terminal box. A typical diagram of a sealing box is shown in Fig. 2.2. The face of the cable sealing box connected

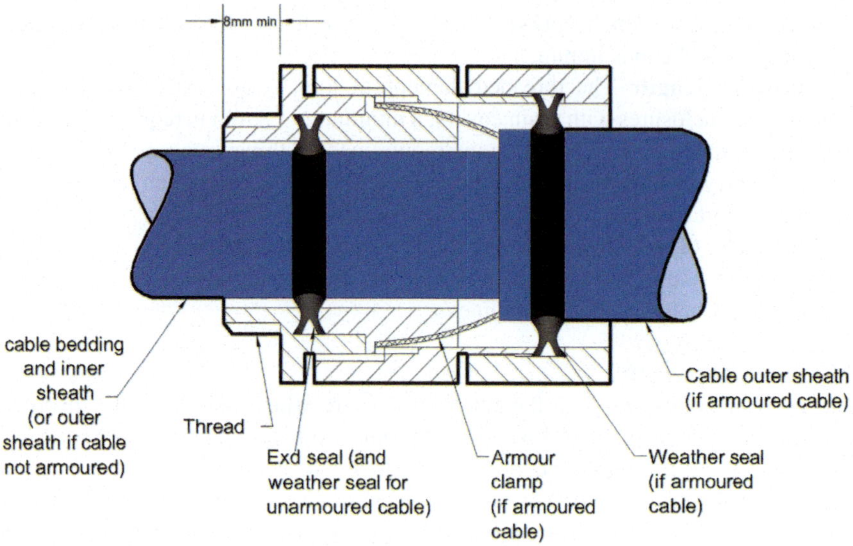

Fig. 2.1 Flameproof compression cable gland

to the terminal box forms a flameproof joint. The sealing box does not allow flame transmission through it when it is sealed with a sealing compound or epoxy resin after passing cables. The cable sealing box is hydraulically tested for leakage at 20 kg/cm^2 pressure for Gr. I applications and 30 kg/cm^2 pressure for Gr. II applications. Cable sealing gives more mechanical strength to the cable to protect it against slippage.

Other details that are covered by the standards are:

- Roughness of the flameproof enclosure should be minimum (machined surfaces).
- Gap between flange joints should be measured by feeler gauzes.

Fig. 2.2 Sealing box

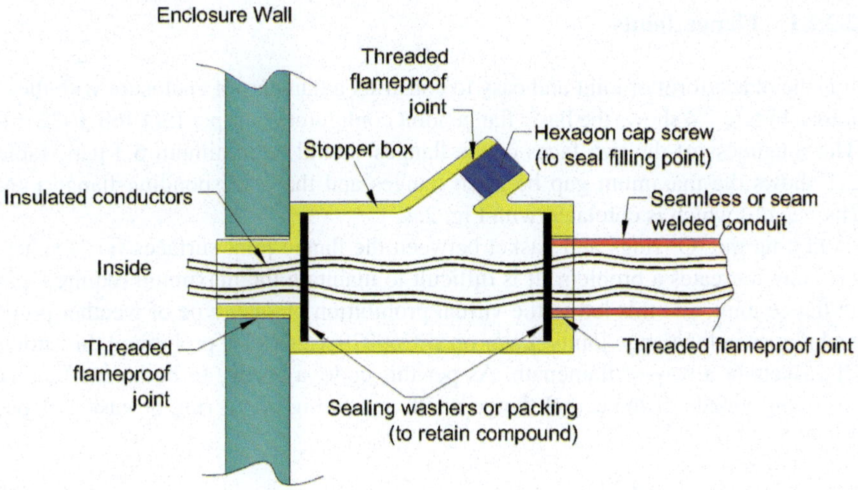

Fig. 2.3 Conduit stopping box

- Holes in joint surfaces as per standard.
- Minimum number of threads engaged.
- Gaskets and 'O' rings.
- Operating rods and spindles.
- Shaft and bearings.
- Breathing and draining devices.
- Mechanical strength, pressure piling.
- Cable gland and cable connection techniques.

Stopping Box
It is a device through which only conductors are passing (i.e., not cable) which is filled with compound. Figure 2.3 shows typical stopping box.

2.2.2 *Joint Between External Atmosphere and Interior of Flameproof Enclosure*

There are five types of joint that are applicable in flameproof enclosures. These are:

1. Flange joints
2. Spigot joints
3. Operating rod joints and cylindrical joints
4. Threaded joints
5. Cemented joints

2.2.2.1 Flange Joints

It is the oldest form of joint and easy to construct a flameproof enclosure with these joints. Figure 2.4 shows the basic flange joint configuration as per ISO 468-1982 [6]. The roughness of the two faces of the flanges should be minimum 6.3 μm. Table 2.1 shows the maximum gap between flanges and the corresponding flange joint (flamepath) which is corelated with Fig. 2.4.

Putting the 'O' rings and gasket between the flange joint surfaces for weather-proofing it creates a problem. It is difficult to maintain the maximum required gap in flange joint, and this led to the virtual prohibition of such type of weather proof techniques in the flange joints. Later on IEC60079-1:2014 [4] prohibited this fitting of gaskets in a way of flamepath. As per this code, a groove to be inserted either inside or outside of the actual flange width and fitting of 'o' ring or gasket as per Fig. 2.5.

Fixings for Parts of Flanged Enclosure

Flameproof equipment farming the flange joint is required fixing, if the fixings are outside the joint then flange width of equipment would be significantly increased. The bolt clearance hole is a passage to the outside environment or the interior of the enclosure. Table 2.2 shows the bolt clearance holes and thread size.

Table 2.3 shows the length of path (Fig. 2.6) between the unprotected bolt hole and interior or exterior of the enclosure.

Bolts passing through the main body of the enclosure from the outside must not adversely affect the flameproof properties of the enclosure. First thing is that if the thread of bolt penetrates the flameproof enclosure and it becomes the part of flamepath, it must comply with the requirement of threaded joints. The flameproof enclosure wall thickness between the threaded hole and within the enclosure (Fig. 2.6) must at least conform to Table 2.4 if the threaded hole does not pass through the enclosure.

There are strict limits on the size of flameproof enclosures with flanged joints for combustible media IIC that use hydrogen and acetylene gases as reference gases.

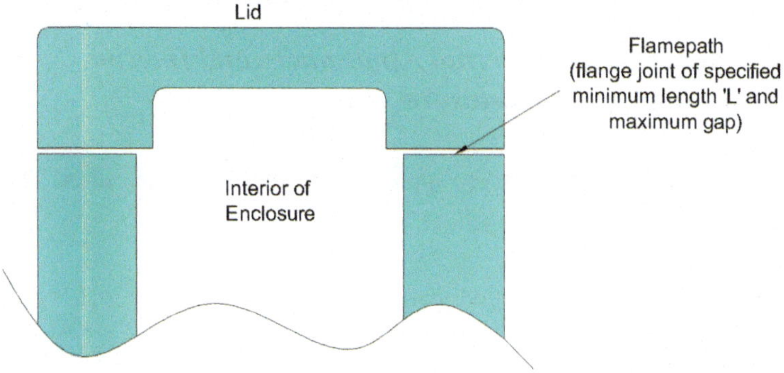

Fig. 2.4 Flange joint in flameproof enclosure

Table 2.1 **a** Minimum width of joint and maximum gap for enclosures of Groups I, IIA, and IIB (IEC60079-1:2014). **b** Minimum width for flange joint and maximum gap for Group IIC enclosure

(a)

Type of joint	Minimum width of joint L (mm)	Maximum gap mm														
		For a volume (cm^3) $V \leq 100$			For a volume (cm^3) $100 < V \leq 500$			For a volume (cm^3) $500 < V \leq 2000$			For a volume (cm^3) $2000 < V \leq 5750$			For a volume (cm^3) $V > 5750$		
		I	IIA	IIB	I	IIA	IIB	I	IIA	IIB	I	IIA	IIB	I	IIA	IIB
Flanged, cylindrical or spigot joints	6	0.30	0.30	0.20	N/P	N/P	N/P	N/P	N/P	N/P	N/P	N/P	N/P	N/P	N/P	N/P
	9.5	0.35	0.30	0.20	0.35	0.30	0.20	0.08	0.08	0.08	N/P	0.08	0.08	N/P	0.08	N/P
	12.5	0.40	0.30	0.20	0.40	0.30	0.20	0.40	0.30	0.20	0.40	0.20	0.15	0.40	0.20	0.15
	25	0.50	0.40	0.20	0.50	0.40	0.20	0.50	0.40	0.20	0.50	0.40	0.20	0.50	0.40	0.20

(b)

Type of joint	Minimum width of joint L (mm)	Maximum gap mm			
		For a volume (cm^3) $V \leq 100$	For a volume (cm^3) $100 < V \leq 500$	For a volume (cm^3) $500 < V \leq 2000$	For a volume (cm^3) $V > 2000$
Flanged joint	6	0.10	N/P	N/P	N/P
	9.5	0.10	0.10	N/P	N/P
	15.8	0.10	0.10	0.04	N/P
	25	0.10	0.10	0.04	0.04

Note N/P is read as not permitted

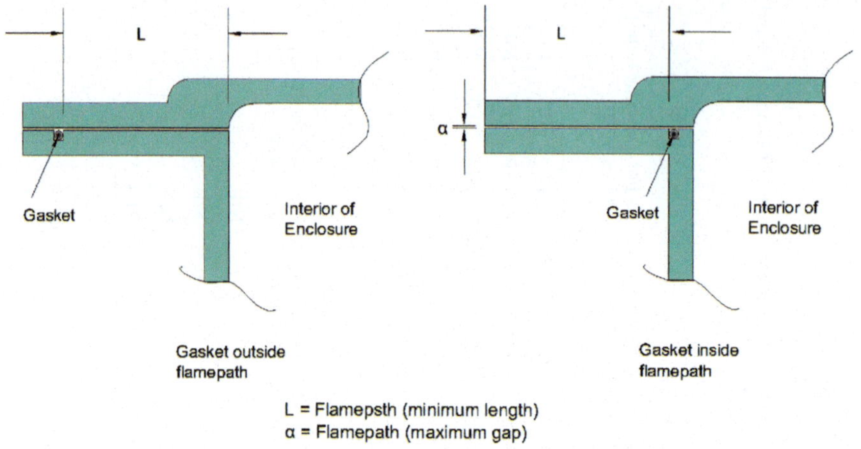

Fig. 2.5 Fitting of gaskets in flanged joint

Table 2.2 Bolt clearance holes diameter with thread size as per IEC60079-0:2017 [5]

Diameter of clearance hole (mm)	Thread size
M24	26
M20	22
M16	18
M14	16
M12	14
M10	11
M8	9
M6	6.6
M5	5.5
M4	4.5

Table 2.3 Joints where bolt crosses the flameproof enclosure

Flamepath (mm)	Length of flamepath between enclosure and exterior (*l*) mm
≥ 25	9
≥ 12.5 < 25	8
< 12.5	6

Enclosures with a volume of more than 500 cm^3 cannot be used in IIC gas groups with flange joints. For IIC gas group, enclosures with a volume of 100 cm^3 must have a flange width greater than 9.5 mm. Flange joint is not permitted in an acetylene explosive atmosphere; acetylene is the reference gas subgroup of IIC. In an acetylene atmosphere, carbon is formed when an explosion occurs. When a flange joint is

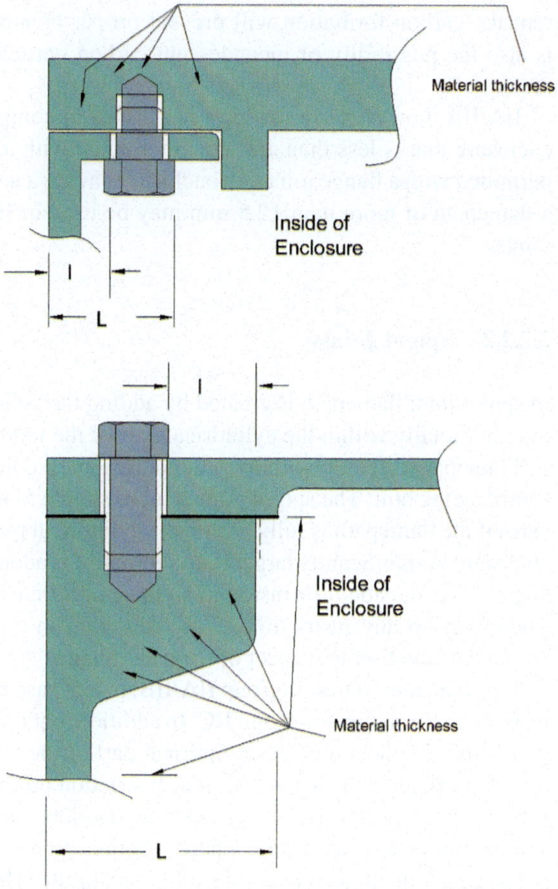

Fig. 2.6 Threaded hole in flameproof enclosure

Table 2.4 Wall thickness between thread and interior of enclosure

Thread size (mm)	Minimum thickness of wall (mm)
M24	8.0
M20	6.7
M16	5.4
M14	4.7
M12	4.0
M10	3.4
M8	3.0
M6	3.0
M5	3.0
M4	3.0

remade, carbon formation will prevent proper breathing. During explosions, there is also the possibility of incandescent carbon particles that may ignite flammable media.

IIA/IIB, however, are the least hazardous in comparison to IIC; in this case, an enclosure that is less than 100 cm^3 in volume with a flamepath of 9.5 mm may be permitted with a flange joint. An enclosure having a volume of less than 500 cm^3 and a flamepath of more than 12.5 mm may be used in IIA/IIB gas groups with flange joints.

2.2.2.2 Spigot Joints

In spigot joint flamepath is created by adding the cylindrical part and flange portion or either totally within the cylindrical part of the joint (Fig. 2.7).

Flamepath is totally within the cylindrical part of the joint so plane part to provide for fixing the bolt. The spigot joint is more simple to manufacture flameproof enclosures if the flamepath is fully within the cylindrical part of the joint. In this situation, enclosure to significantly larger if its alternative is adopted. This type of joint has right angle either outside it or inside it, so there is little chance to emit ignition particle. Therefore, no any restriction to use this joint in acetylene explosive atmosphere unlike the situation in respect of their flat flanged counterparts.

Equipment used in sub-group IIA/IIB, in that case plain part of the joint must be at least 3 and 6 mm for sub-group IIC. In addition, equipment used in IIC gas sub-group then minimum length of the cylindrical part of the joint must be at least half of the minimum flamepath. No any similar restriction applies for the enclosures used for sub-group IIA or IIB. There is no specific restriction to the use of flat flange joint and spigot joint in acetylene atmosphere. There is no restriction on the minimum length of flange width allowed in combustible media IIC. The following restrictions apply:

1. The flamepath less than 12.5 mm (cylindrical part plus plane part) is not permitted for sub-group IIC in spigot joint.
2. The flamepath less than 25 mm (cylindrical part plus plane part) of enclosures with volume greater than 2000 cm^3 is not permitted for sub-group IIC in spigot joint.
3. The flamepath less than 9.5 mm (cylindrical part plus plane part) of enclosure with volume greater than 100 cm^3 is not permitted for sub-group IIA/IIB in spigot joint.
4. The flamepath less than 12.5 mm (cylindrical part plus plane part) of enclosure with volume greater than 500 cm^3 is not permitted for sub-group IIA/IIB in spigot joint.

These requirements are more severe than those applied to spigot joints where total flamepath covered by cylindrical part of the joint.

Gaskets in Spigot Joint for Weatherproofing

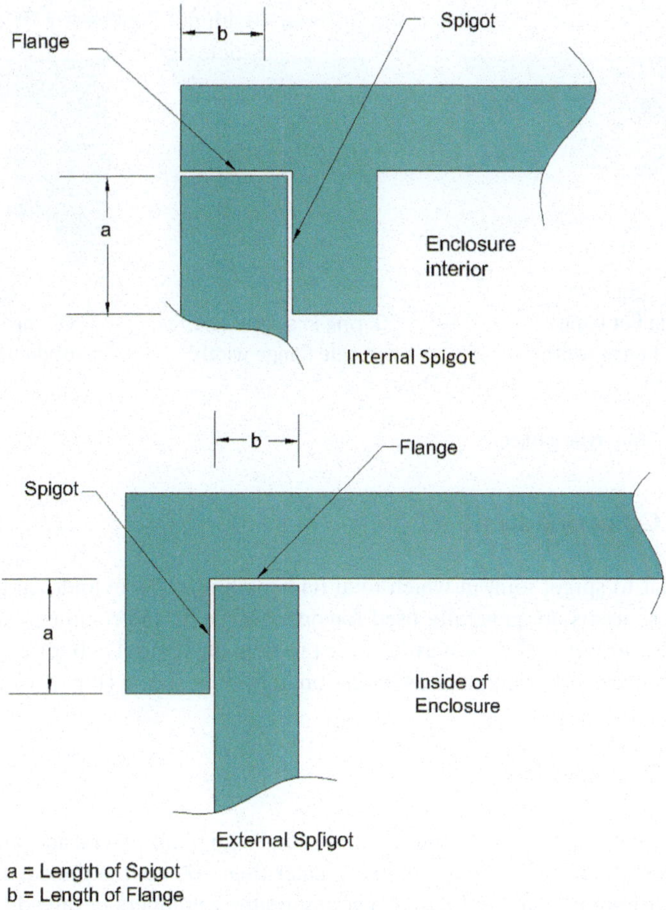

Fig. 2.7 Internal and external spigot joints

Figure 2.8a shows the 'o' ring-type gasket in the plane part and the cylindrical part of the joint.

In Fig. 2.8b, c 'o' ring gasket in both inside the flamepath and outside the flamepath. 'o' ring gasket must always be inside the flamepath or outside the flamepath and should not within it.

Figure 2.8, the flamepath must be totally within the cylindrical part where the 'o' ring is outside the flange width but in the cylindrical part of joint. If 'o' ring is in the plane part of joint outside the flamepath in this case composite joints are possible.

The flat gaskets are also permitted either inside or outside the flamepath as shown in Fig. 2.9. In this case, flamepath must be totally in cylindrical path.

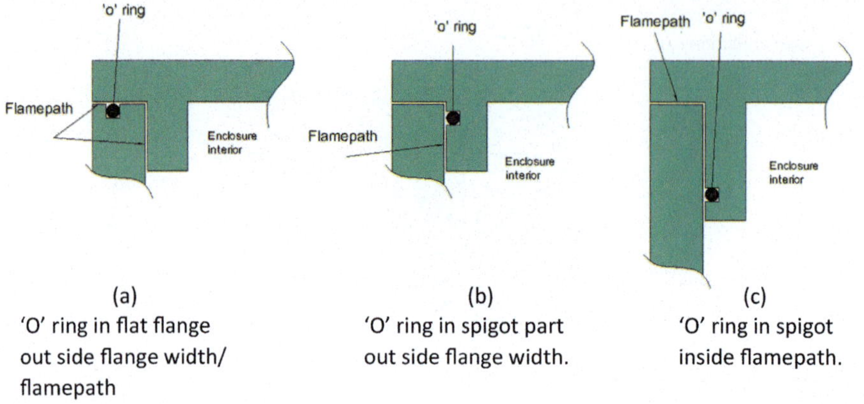

Fig. 2.8 'o' ring-type gasket

2.2.2.3 Cylindrical Joint

It is similar to spigot joint in which total flamepath is in the cylindrical part of the joint. These joints are generally used for operating rods and rotating shafts where mainly eccentricity and wear may become problematic. Partial cylindrical joints are permitted where plates are used as covers on hollow cylinders (Fig. 2.10).

2.2.2.4 Threaded Joints

It is commonly used for cable entries, enclosures with plain covers, and small flameproof boxes. Enclosure having volume greater than 100 cc, the threaded length of flameproof joint is minimum 8 mm. The minimum 5 numbers of threads should be engaged and pitch size may be between 0.7 and 2 mm for parallel threads and at least 0.9 mm pitch size for tapered threads.

2.2.2.5 Cemented Joint

It is often used to insert viewing parts into flameproof enclosures, and it is difficult to create a joint which complying the requirement of flameproof joint. In such cases, a joint may be made using cement and this joint is always shorter than mentioned flameproof joint. Cement may be generally used for fixing glass windows in flameproof enclosure. The strength of the enclosure does not depend upon cement or sealing material. It is desired to have a backing ring over and above the cement. Cemented path as per IEC60079-1:2014 is mentioned in Table 2.5.

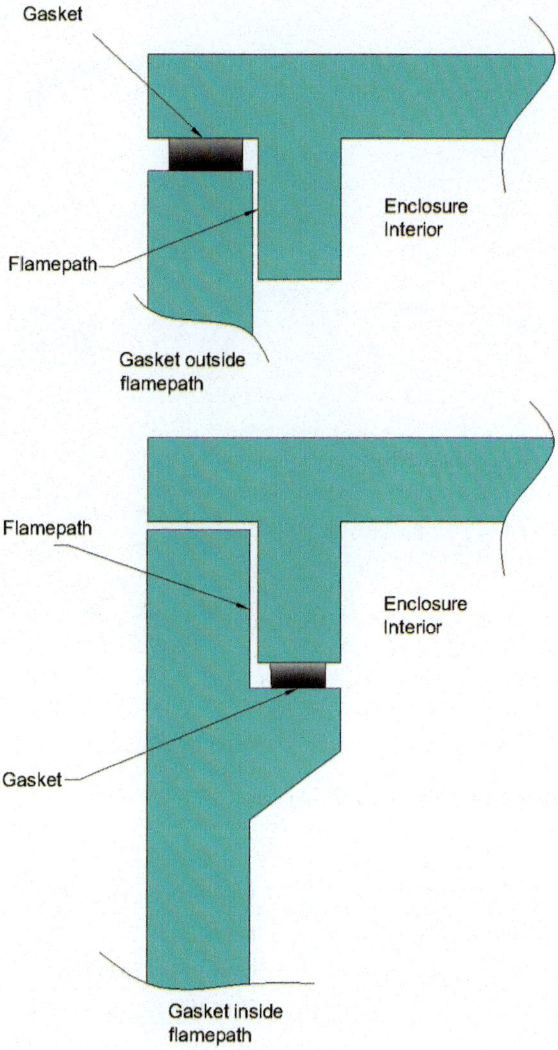

Fig. 2.9 Flat compressible gasket, arrangement in flameproof enclosure

2.2.2.6 Labyrinth Joint

Labyrinth joint is similar to cylindrical joint (Fig. 2.11).

In labyrinth joint, no maximum limit of radial clearance is specified. During testing, only flame transmission test should be passed. The sleeve bearing with a rotating machine is only applicable in sub-group IIA/IIB. Sleeve bearing is not applicable in gas group IIC atmosphere.

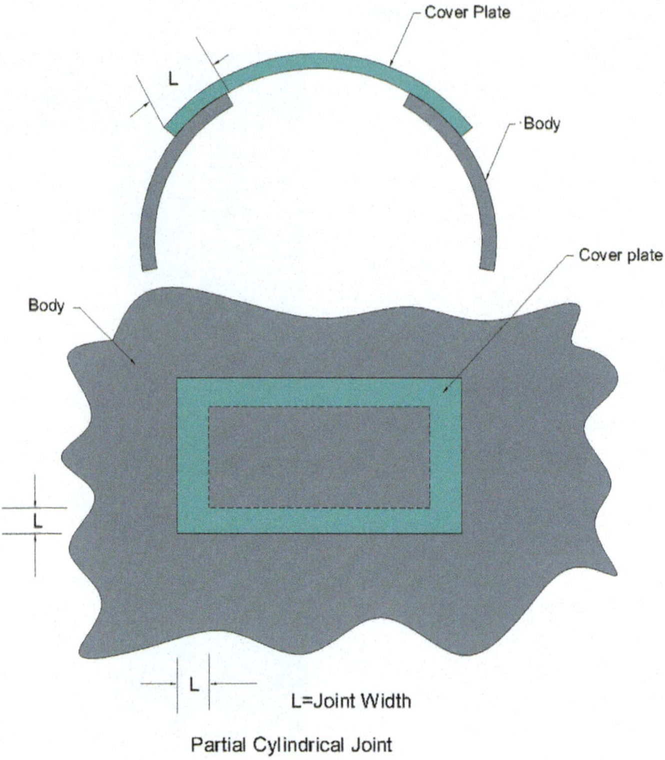

Fig. 2.10 Partial cylindrical joint

Table 2.5 Minimum length of cemented joint with respect to volume of enclosure

Volume of enclosure (cc)	Cemented length (mm)
Less than 10	Al least 3
Over 10 and up to 100	At least 6
Over 100	At least 10

2.2.2.7 Joints for Rotating Machines and Shafts

A cylindrical or labyrinth flameproof joint is used in a rotating electrical machine with sleeve bearings. At least one face of the joint shall be made by lead brass of non-sparking metal. The air gap between stator and rotor is greater than the minimum radial clearance K.

d shaft diameter
k minimum radial clearance permissible without rubbing
m maximum radial clearance
$(D - d)$ diametrical clearance

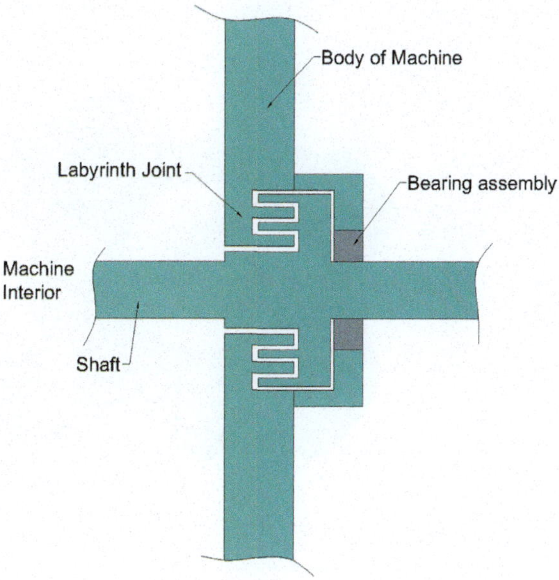

Fig. 2.11 Geometry of labyrinth joint

A rotating shaft is a unique case due to rotation. Flame proofness is difficult to maintain. The joint length must be equal to the shaft diameter if that is below 25 mm. It is not necessary to make the joint longer than 25 mm. The minimum value of radial clearance (k) in joint should be 0.05 mm, and this joint is only applicable in gas group IIA and IIB. In case of labyrinth joints the same radial clearance is required. For a rotating shaft, flat flanged and cylindrical joints can be used as floating joints. Such joints are not suitable for gas group IIC atmospheres (Fig. 2.12).

Greasing Arrangements for Bearing
Grease groove arrangements are made in shafts to retain grease around the bearing, and these should not interfere with the flamepath and gap (Fig. 2.13).

2.2.3 Construction of the Flameproof Enclosure

Flameproof products are made from metals and non-metals (usually plastic), but in India non-metal materials are not allowed for construction. Non-metal products are limited to volumes not more than 3000 cm^3. Generally, cast iron, brass, and light metal alloys are used for the construction of flameproof enclosures. After the enclosure is constructed by fabrication, it is tested four times the reference pressure. The enclosure may be constructed by molding or fabrication. Non-metallic enclosures having non-metallic joint surfaces are subjected to a flame erosion test. This test is carried out by conducting 50 tests in the enclosure. The comparative tracking index (CTI) value of a non-metal enclosure should be known. As per IEC 60112:1979 [7], both metallic

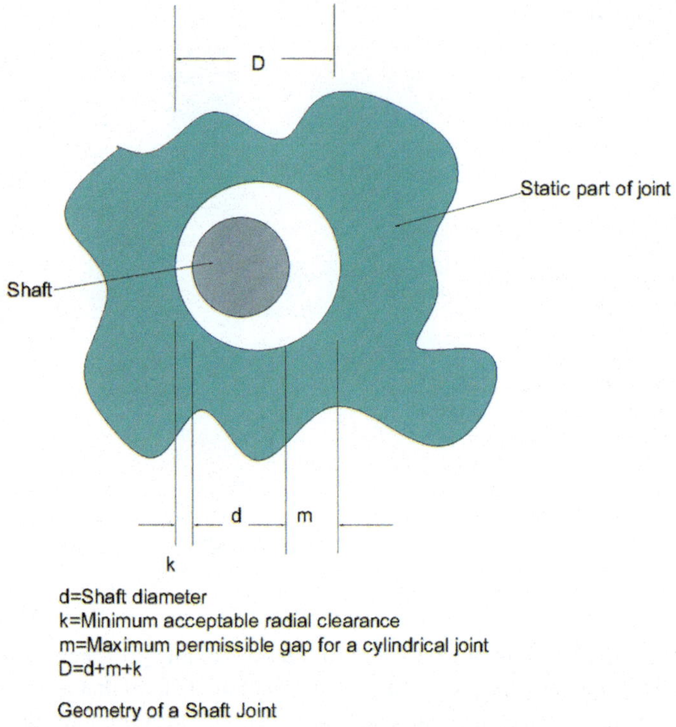

Fig. 2.12 Joints of shaft glands of rotating electrical machine

and non-metallic enclosures should be subjected to hydraulic pressure test at 1.5 times the reference pressure. In pressure tests, enclosures may move elastically, but permanent distortion is not acceptable. In order to confirm this, the enclosure must be measured after the pressure test. Flarepath, gap, or enclosure parts that are permanently distorted are unacceptable. If there is any elastic expansion occurs in the flamepath, then it should be checked during the external ignition test. The reference pressure test is carried out to check the mechanical integrity of the product. During testing, a specific test mixture is ignited inside the enclosure, and the pressure is monitored as it rises and falls. At least five tests are carried out, and the worst-case pressure is taken as the reference pressure. The maximum pressure developed inside the flameproof enclosure is called 'Reference Pressure' (Fig. 2.14).

Pressure Pilling

A pressure piling phenomena occur in the enclosure is divided into separate parts as per Fig. 2.15. Pressure piling is a condition resulting from ignition of pre-compressed gases in compartments other than those in which ignition was initiated.

Figures 2.16 and 2.17 show the indirect and direct entry in flameproof enclosure respectively. Earlier Bureau of Indian Standard (BIS), code IS 2148:2081, was not

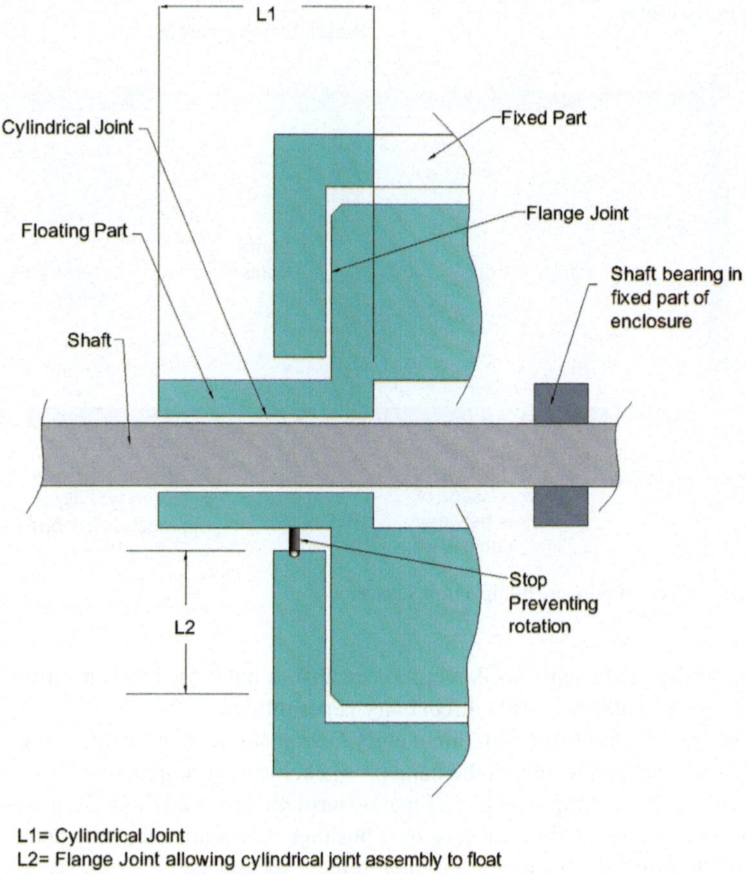

Fig. 2.13 Dimension for shaft joint of rotating machine

Fig. 2.14 Variation of explosion pressure with time

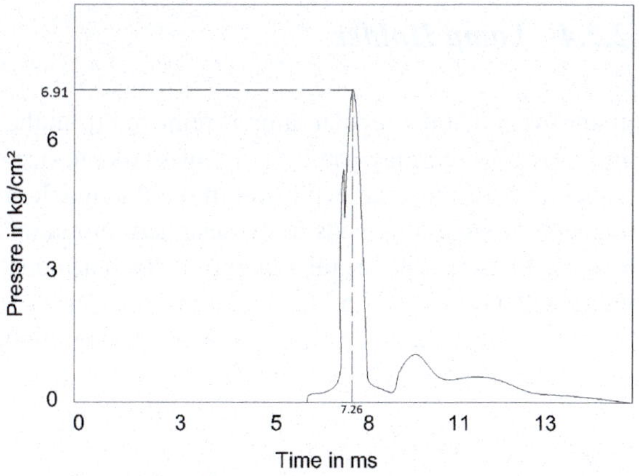

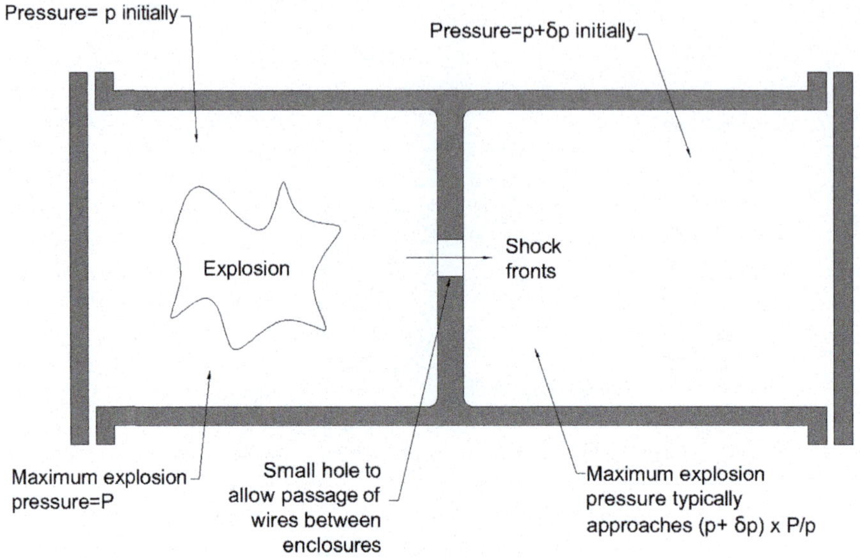

Fig. 2.15 Effects of pressure piling during explosion

permitted the direct entry in flameproof enclosure but after harmonization of this code as per IEC60079-1:2004 direct entry is permitted.

Flameproof equipment with direct entry having connection facilities, e.g., terminals for external connection in the flameproof enclosure. Equipment with an indirect entry enclosure of equipment has a separate terminal box, and this terminal box may be connected to the main enclosure by a bushing. The main enclosure and terminal box should fulfill the flameproofing requirement separately.

2.2.4 Lamp Holder

In the event that the regular lamp is removed from the lamp holder, it becomes loose in service, which may cause an arc of sparks to occur in an explosive atmosphere, which may result in an explosion in the flammable media.dia. A mechanism like a flameproof situation needs to be designed into lamps so that in the event sparking occurs, the lamp will be either intentionally or accidentally cut off. Figure 2.18 shows that smaller sizes (E_{10} and E_{14}) require that at least two complete threads of the lamp must be engaged at the moment of contact separation.

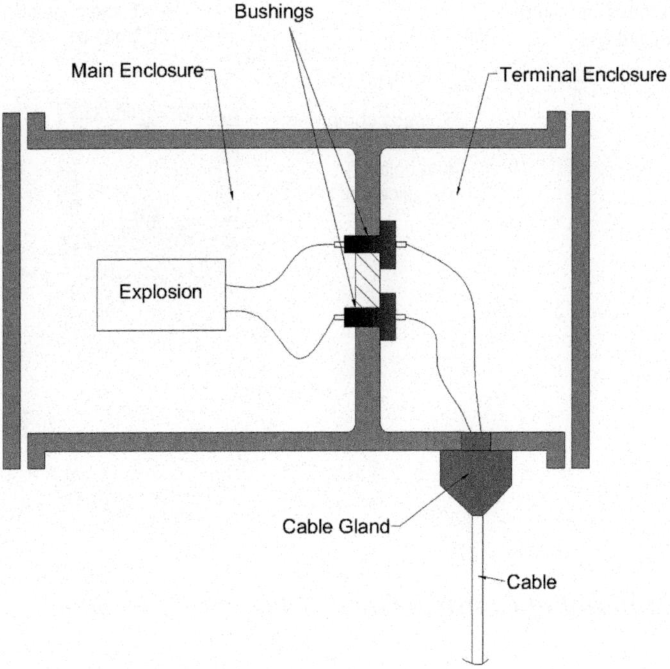

Fig. 2.16 Flameproof enclosure with indirect entry

Fig. 2.17 Flameproof enclosure with direct entry

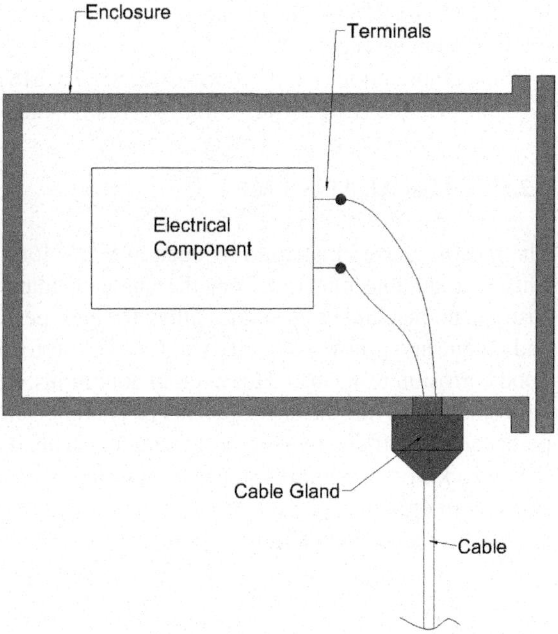

Fig. 2.18 Lamps and lamp caps for E_{10} and E_{14}

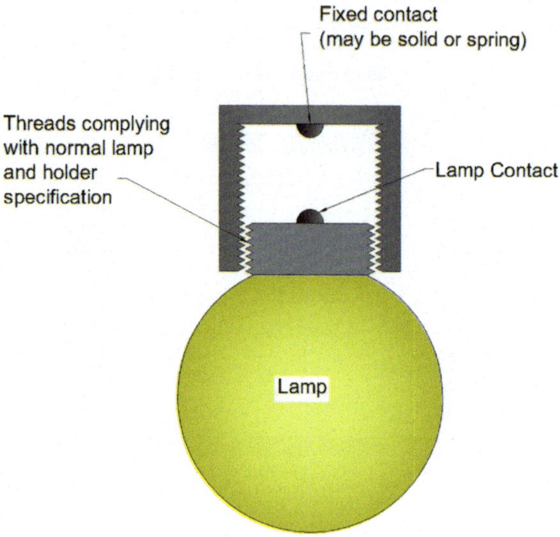

2.2.5 Material of Construction of Flameproof Product

Following material of construction is generally specified:

a. Cast Al. Alloy LM6
b. SS 304 and 316 (welded plate construction)
c. Cast stainless steel
d. Sheet fabrication (Al. Alloy, SS 304, SS316, MS)
e. Non-metallic (FRP, SMC, DMC, polycarbonate)

2.2.5.1 Cast Al. Alloy LM6

The most preferred material for junction boxes for surface industries is cast alloy. This is a suitable material from the angle of manufacturing and uses both. The casting and machining properties allow for all types of construction (flange, spigot, and tongue and groove) as well as all types of protection (Exd, Exe, and WP). It has good corrosion resistance. However, in conditions where there is a highly corrosive atmosphere like marine environments or alkaline/sulfurous fumes, the junction box paint can peel off due to white rust formation. This is more due to incorrect painting than a lack of corrosion resistance of aluminum alloy. In such an environment, the flameproof enclosure should be used with the proper grade of paint or even better; it should have an anodized finish or zinc spray finish.

2.2.5.2 SS 304 and 316 (Welded Plate Construction)

For field conditions with highly corrosive atmospheres such as marine environments or alkaline/sulfurous fumes, SS 304 and 316 constructions are used. However, due to the difficulty of casting and machining stainless steel, it is practical to go for welded plate construction in the case of Ex d junction boxes. This is because it is required in SS construction. Since stainless steel has a higher density than aluminum alloy and because it costs more to manufacture, welded stainless steel junction boxes are usually 6 times more expensive than an aluminum alloy enclosure with a similar flameproof rating. Epoxy powder coating is a recommended finish to provide additional corrosion protection.

2.2.5.3 Cast Stainless Steel

During the casting of SS, case hardening is taking place. Cast SS needs to be normalized to improve machinability. In addition, machining cast SS is a difficult and costly exercise. Due to these limitations, only a small junction box in a round body with spigot-type joints or threaded joints is practical in Exd type. As a result, junction boxes in cast steel construction should only be specified for smaller sizes in round bodies with spigots or threaded joints. Epoxy powder coating is a recommended finish to provide additional corrosion protection.

2.2.5.4 Sheet Fabricated (Aluminum Alloy, SS 304, SS316, MS)

Sheet steel construction is inexpensive and suitable for mass production. However, due to thickness limitations, sheet fabrication can only be used for Exe enclosures and not for Exd flameproof enclosures. Therefore, Exd flameproof enclosures must be cast or welded. The sheet-fabricated flameproof enclosure can be constructed in various materials, such as Aluminum Alloy, Stainless Steel 304, SS316, and MS to suit corrosive, marine, and non-corrosive environments. Epoxy powder coating is recommended finish to provide additional corrosion protection. However, for aluminum alloy, an anodized or zinc spray finish is recommended.

2.2.5.5 Non-Metallic (GRP, FRP, SMC, DMC, Polycarbonate)

Non-metallic junction boxes are available in a variety of materials, including fiber reinforced polyester (FRP), GRP (glass reinforced polyester), sheet molding compound (SMC), dough molding compound (DMC), and polycarbonate. In case of FRP instead of glass fiber, other fibers like carbon may be used. Generally, both GRP and FRP are made of glass fiber. Basic constituent of GRP/FRP/SMC and DMC are same, i.e., polyester and glass fiber. The difference being that GRP of FRP cures at room temperature and atmospheric pressure when mixed with accelerator and hardener, whereas in case of SMC and DMC curing takes place at high temperature and pressure. SMC and DMC differ in that they are available as sheet and dough, respectively in the semi-finished stage.

The epoxy spray paint finish is recommended for FRP/GRP JBs. SMC/DMC are molded finishes as pigments are blended during the molding process.

Polycarbonate is used for small junction boxes only due to poor impact resistance. Carbon fiber is blended in polycarbonate to improve its impact resistance. Polycarbonate has a molded finish since the pigment is blended during the molding process.

Non-metallic junction boxes are available in type Ex e due to code restrictions. These are preferred in highly corrosive atmospheres due to their good corrosion resistance. These are suitable for installation on cables without armor. However, these are generally less suitable vis-a-vis metallic junction boxes due to many reasons like:

- Poor impact resistance—Prone to breakage/damage.
- Poor UV resistance—Prone to aging and losing mechanical strength due to sunlight.
- Flammable—Prone to damage in case of fire and creates a smoke hazard.
- Organic vapors can cause pitting and chemical damage due to chemical reaction.
- Earthing bolts/hardware prone to accelerated corrosion.

2.2.6 Test for a Flameproof Apparatus

2.2.6.1 Design Evaluation and Physical Examination

Flameproof or explosion-proof enclosures are strong enough to withstand the pressure generated inside and prevent flame transmission from inside to outer explosive environments. The Flamepath and gap between metal to metal and metal to non-metal of any flameproof enclosure use complied with code IEC60079-1:2014. Non-metal to non-metal flamepath is not permitted by this code. External connection to the apparatus should normally be indirect, i.e., through an integral terminal box or non-integral terminal box. A cable entry device such as a compression-type cable gland should be independently explosion-proof. Flame should not travel through the gap between the non-metallic cable sheath and the neoprene ring used for compression. Glass windows and glass covers should be toughened and properly sealed to the metal housing in order to prevent the flame from passing through. An apparatus's separate housing should be individually flameproof, i.e., explosions should not travel between enclosures through cable entry devices.

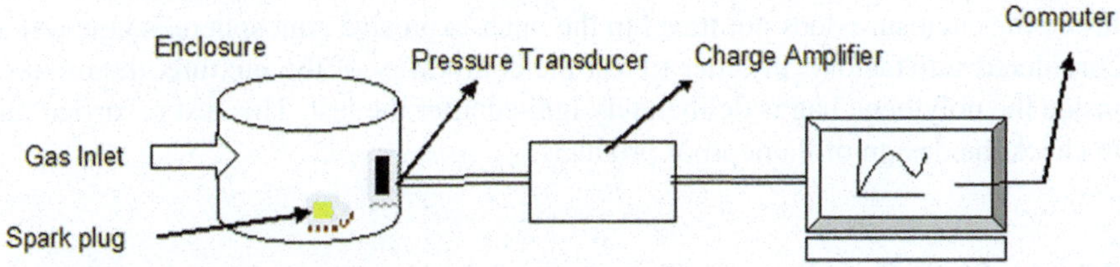

Fig. 2.19 Dynamic explosion pressure recording system

Fig. 2.20 Variation of dynamic pressure with time

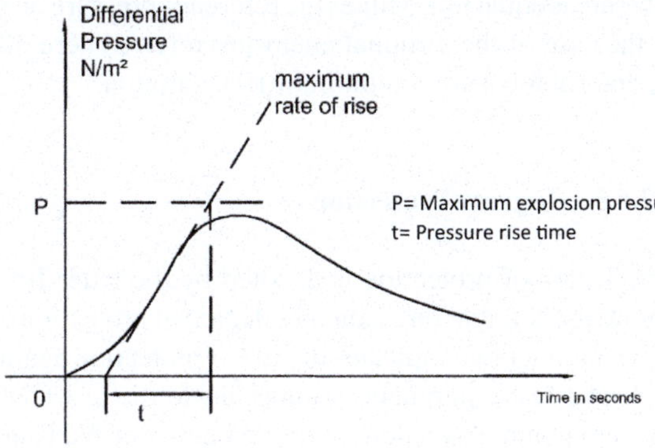

2.2.6.2 Determination of Explosion Pressure

Explosion Pressure Test

The tests are carried out by igniting the explosive gas mixture inside the flameproof enclosure. The maximum pressure developed inside the enclosure is measured by a piezoelectric pressure transducer. This maximum pressure developed inside the enclosure is called the reference pressure. This test involves changing the position of the pressure sensor and spark plug five times. The mechanical integrity of this enclosure is checked by this test. The reference pressure depends upon so many parameters like humidity in air, orientation of source of ignition, energy of spark, release of burnt gases through gaps, and also internal content of the enclosure. Therefore, it is not expected that all the parameters will be identical during test. As a result, the standard requires us to make the test 5 times and use the highest value as a reference pressure (Figs. 2.19 and 2.20).

2.2.6.3 Non-transmission of an Internal Ignition

In the non-transmission of internal ignition test, a flameproof enclosure is placed inside a polythene bag and the same flammable mixture is placed inside the enclosure and the polyethylene chamber. The flammable mixture inside the enclosure is ignited by a spark plug. Any flame or combustible product generated during the explosion inside the enclosure does not travel to the outer explosive atmosphere, so the test is considered satisfactory. In order to test the conformity of the mixture, the mixture inside the polythene bag is deliberately ignited after the test. This test is carried out to check the design of flameproof products.

2.2.6.4 Static Over Pressure Test

The enclosure should be subjected to a static pressure test, such that the applied pressure is equal to 1.5 times the reference pressure or 3.5 bar, whichever is maximum. In the case of the terminal enclosure referred here, the static overpressure test is to be done at 8 bar for a minimum 10 s duration.

2.2.6.5 Ingress Protection

The degree of protection is denoted by the letter IP followed by two characteristic numerals. Number one indicates degree of protection against ingress of solid particle, while number two indicates degree of protection against ingress of liquid (Fig. 2.21).

Under-test equipment is hung inside the test chamber which is connected to a vacuum pump. A vacuum of around 200 mm Wc is maintained inside the enclosure. This test chamber was filled with 2 kg of talcum powder per m^3, and talcum powder is kept in suspension by air current. This condition was maintained for two continuous hours during which total air was drawn through the joints. Table 2.6 shows the degree of ingress protection (IP) with respect to dust and liquid.

Fig. 2.21 Ingress protection test setup

Table 2.6 Numerals related to degree of ingress protection (IP-XX)

	First number		Second number
0	No protection	0	No protection
1	Objects greater than 50 mm	1	Vertical dripping water
2	Objects greater than 12 mm	2	75–90 °C dripping water
3	Objects greater than 2.5 mm	3	Sprayed water
4	Objects greater than 1 mm	4	Splashed water
5	Dust protected	5	Water jets
6	Dust tight	6	Powerful water jets
		7	Effects of immersion
		8	Indefinite immersion

2.2.6.6 Surface Temperature Rise Classification

The electrical equipment due to the resistance of conductors and intense radiation of light becomes hot. If flammable gases exist in the surrounding area, the surface temperature of the body or glass part should not exceed the auto-ignition temperature. The temperature classification of a flameproof motor is conducted by running it at rated voltage and rated current. The maximum surface temperature observed at any location after equilibrium is considered for T-classification as per IEC60079-:2014. The temperature class and maximum temperature are given here for reference (Table 2.7).

According to IEC60079-1:2014, in underground coal mines, the maximum surface temperature allowed for flameproof equipment is 150 °C, where coal dust can form a layer. The maximum surface temperature should not exceed 450 °C, where coal dust is not expected to form a layer. In later case the maximum surface temperature should be marked on the apparatus or 'X' symbol is placed after certificate reference to indicate the conditions for safe use.

Table 2.7 Temperature class and corresponding maximum temperature

T-class	Maximum surface temperature in °C
T6	85
T5	100
T4	135
T3	200
T2	300
T1	450

2.2.6.7 Impact Test of Electrical Apparatus

The height of fall for various impact energy requirement may be estimated from the relationship.

$$h = E/mg$$

where h—height in meter, E—impact energy in Joule, m is mass in kg, and $g = 10$ m/ s^2.

The impact energy required for Gr. I enclosure is 20 J, while for Gr. II enclosure, it is 7 J only. Similarly, for glass parts it is 7 J and 4 J respectively for use in Gas Gr. I and Gas Gr. II applications respectively.

2.2.6.8 Thermal Shock Test

When performing thermal shock testing of glass that is used for an inspection window or for lighting fixtures, the maximum surface temperature at the rated supply is considered. The glass is kept in the oven at the maximum service temperature for 15–20 min, and then a water jet of 100 °C is sprayed on it. Glass should not be cracked or damaged during this process.

2.2.6.9 Healthiness Assessment of Flameproof Apparatus

As per code requirements, flameproof equipment made of cast iron (150, MS or SS) is used in underground coal mines. During coal mining, the atmosphere is corrosive and rust is a major contributor to the degradation of flameproof enclosures. Due to rust, enclosure bolts become punctured, which may affect the flameproofness of the enclosure. It may be minimized by using anti-rust paint at regular intervals. Due to roof falls and falling stones, coal mine conditions are very absurd. So the probability of loosening the bolt of the flameproof enclosure. Therefore, bolts are provided with shrouded and spring washers to prevent any loosening in service.

In underground coal mines, coal dust is also ignitable. Therefore, electrical equipment used in coal mines should be dust tight. Coal dust accumulated on the surface of the flameproof enclosure, so it violated the temperature classification because coal dust is a bad conductor of heat [8].

Improper cable termination in flameproof equipment may disturb the flameproofness of the equipment. The sealing ring in the cable gland maintains the flamepath with the cable passing through it when compressed in a gland. During cable gland maintenance, a single sealing ring cannot be used for different cable diameters. Before terminating cables in terminal boxes, they should be sealed in cable sealing boxes, otherwise terminal boxes are not flameproof, since unsealed sealing boxes allow flame to travel to outer flammable environments. Unplugged cables entries should be properly plugged. All holes are closed with stopping plugs during service. These

Table 2.8 Impact energy for glass part used in the U/G coal mines

Part name	High impact energy (J)	Low impact energy (J)
Light-transmitting part with guard	4	2
Light-transmitting part without guard	7	4

holes should be properly checked. Generally, wire guards are used to protect lighting fixtures during maintenance. There is change to overlook the wire guard, and it is left unfixed with fitting. This makes the fitting dangerous to use in an explosive atmosphere. Table 2.8 shows impact energy (in Joule) for glass used in the underground coal mines.

Temperature classification of any flameproof product is an essential parameter. It is based on the full rating given by the manufacturer, but the rating of the bulb may change during maintenance. Therefore, its surface temperature is likely to be higher than the auto-ignition temperature of surrounding gases. So it may cause an explosion in a hazardous area.

Manufacturers sometimes add reflectors to their light fittings for specific temperature classes to increase the intensity of light.

2.2.6.10 Summary

Flameproof products testing, certification, and manufacturing are great jobs because they serve as safety equipment in hazardous areas. In addition, they are directly related to the production of mines, minerals, petroleum products, and the lives of people. Failure of equipment in the industry will put people's lives at risk.

User Responsibility
The end user is totally responsible for the safety of the plant. Maintaining flameproof products in such a way as to ensure their safety during installation and maintenance.

Installation

- Installation of flameproof products that have been certified by testing authorities.
- Prevent the occurrence of short circuit currents of dangerous values inside the apparatus.

Use

- Always use flameproof products as certified by testing authorities.
- Equipment should be installed in accordance with certified ratings.
- Do not leave oil-immersed starting resistors in the circuit.

- Limit the temperature rise of the apparatus.

Maintenance

- Clearance and creepage should be maintained.

Keep all joint surfaces clean.

- All bolts, screws, studs, and nuts should be tight.

2.3 Intrinsic Safety Ex 'i'

2.3.1 Introduction

Flameproof protection techniques have been discussed so far. Out of all the above-discussed protection techniques, Flameproof (Ex d) enclosures are the oldest and the most popular for installation in a hazardous area. Flameproof equipment is used in zone 1 and zone 2 areas but not allowed in zone 0 areas where the continuous presence of hazardous gases is observed. Intrinsic safety is the safest method to ensure that electrical equipment is installed in, zone 0, zone 1, and zone 2 hazardous areas and cannot reset in fire or explosion. Intrinsic safety is the second oldest protection scheme after Flameproof [9].

2.3.2 Principles of Intrinsic Safety

'Intrinsic Safety is the type of protection based on the restriction of electrical energy within equipment and of interconnecting wiring exposed to the explosive atmosphere to a level below that which can cause ignition by either sparking or heating effect.' [9].

This type of protection is applicable to electrical or electronic circuits in which the circuit itself is incapable of causing an explosion. This means low-energy devices that operate on low voltage and current.

Example: Audiovisual alarms, cap lamps, mine communication systems, portable gas detectors, small power supplies, hooters, signaling systems, etc.

Two basic criteria to be satisfied in the I.S. circuit by its definition:

A. Restricted energy output (low spark energy).
B. Temperature must not increase (T-Class)

To satisfy these two basic criteria of Intrinsic Safety, we must understand the main cause of the spark and the high temperature in the circuit. According to Ohm's Law,

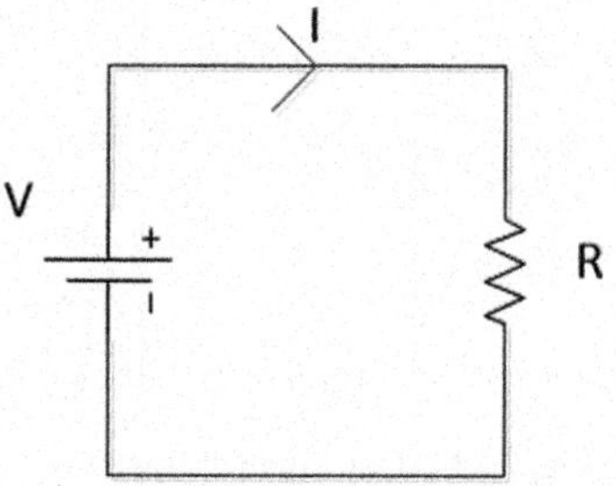

Fig. 2.22 Simple electric circuit with resistive load

the current drawn by the load is directly dependent on the voltage drop across it, i.e., $V = I * R$. Power consumption by resistance is denoted by $P = V * I$ or $P = I2 * R$. This power consumption is the sum of utilized power and lost power. The lost power is mainly in the form of heat energy which is responsible for the temperature rise in the circuit (Fig. 2.22).

2.3.2.1 Causes of Spark in Electrical Circuits

(i) Short circuit current
(ii) Open circuit voltage
(iii) Instantaneous transition of voltage and current
(iv) Energy storage elements

Short circuit current: When the internal resistance of the source is negligible and load resistance becomes near zero, this condition in the circuit is known as a short circuit. During this condition, high current flows are called short circuit currents. In case of short circuits, the wires are not capable of handling such a high current, causing them to melt rapidly. Sparks are produced by this rapid occurrence of short circuits [10]. To avoid such a phenomenon, a fuse of appropriate rating should always be used. A current limiting resistor can also be used to prevent short circuit current (Fig. 2.23).

Open circuit voltage: Voltage across two terminals and wires is responsible for the insulation/air breakdown. It takes 21 kV of voltage to break down 1 cm separated terminal in air. When there is no load at the output terminal and the power supply

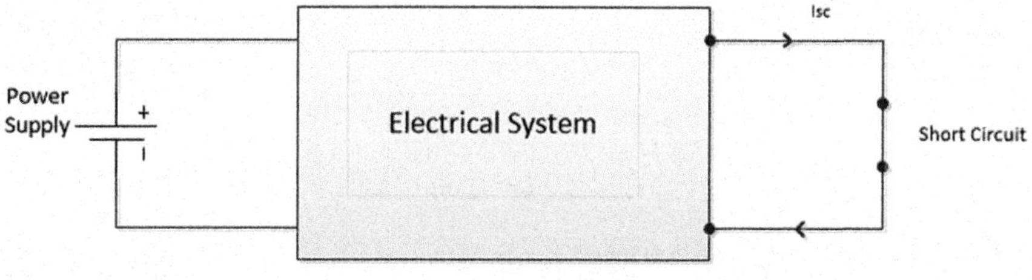

Fig. 2.23 Output short circuit

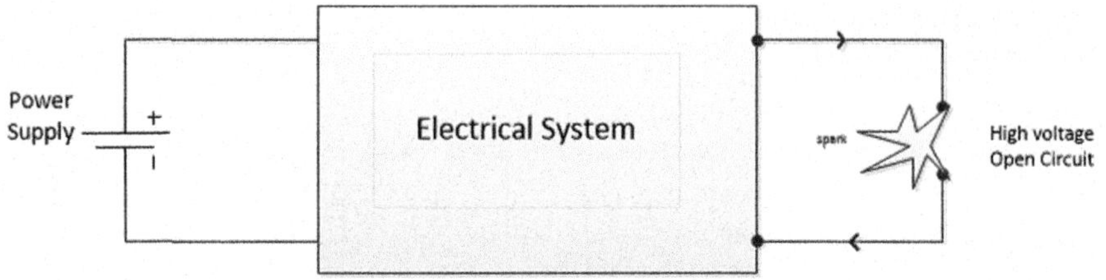

Fig. 2.24 Open circuit output

is on, then the source voltage appears at the output terminal (which is open). This appeared voltage is called open circuit voltage. If the separation/insulation between two terminals of the output is not sufficient or very low, breakdown of the air between them occurs which results in a spark. It is advisable to have sufficient separation (creepage and clearance) between two conducting parts [11] (Fig. 2.24).

Instantaneous transition of voltage and current: Instantaneous transition of voltage and current across the two terminals leads to a high rate of change of current and voltage. A High rate of change of current and voltage from short circuit to open circuit leads to a spark in the circuit. These sparks dissipate high energy in a very small period of time. These occur mainly when a switch is operated from short to open (turning off). We suggest reducing the over voltage (V_{oc}) and short circuiting to control these big sparks (Fig. 2.25).

Energy storage elements: Batteries, inductors, and capacitors are the three main elements which store energy. Inductance and capacitance are the natural effect in the circuit even if physically indoor/or capacitors are not present.

Stored energy is calculated as

$$\text{In indicator } E = 1/2\,\text{LI2}$$
$$\text{In capacitor } E = 1/2\,\text{CV2}$$

Stored energy in the inductor may lead to a high voltage in the circuit if it is somehow kept open or removed from the circuit. VL = L di/dt, so when the inductor is abruptly removed from an energized circuit, there is a sudden change in current. A

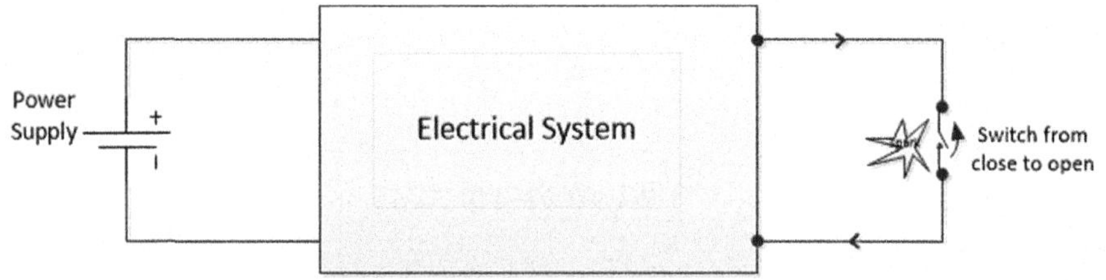

Fig. 2.25 Fast switching effect in output

sudden change in current (high rate of change of current) will result in high voltage and a big spark [12].

When the capacitor is somehow shorted, the stored energy could lead to a high short circuit current. Since ic = C dvc/dt, there is a sudden change in voltage in the circuit when it gets shorted in energized condition. The sudden change of voltage (high rate of change of voltage) will result in a high current, and along with the failure of the capacitor, a high spark will occur.

In the circuit, a protection scheme is applied to inductors and capacitors in order to prevent such conditions. In summary, all four possibilities of spark can be accounted for by energy limiting. Energy limitations as per IEC60079-11:2011 are discussed in Sect. 2.3.3.

2.3.2.2 Cause of Temperature Rise in Electronic Circuits

As the circuits operate, all the components dissipate some energy as heat. High temperatures result from the dissipation of more energy. If the temperature of that component is more than the auto-ignition temperature of the gas present in the surroundings, it will lead to a big explosion.

IEC 60079-11 provides a clear guideline for designing circuit parameters in such a way that temperature rise of a circuit is maintained during normal operation. The methods for determining temperature rise are discussed.

2.3.3 Methods to Avoid Spark and Hot Surfaces

It is necessary to comply with the standard IEC60079-11 to be intrinsically safe. Technically, the main objective is to avoid sparks and hot surfaces, so that ignition is minimized.

An energy-limiting principle is used to prevent sparks. The energy-limiting principle is primarily concerned with the values of voltage and current in circuits that are exposed to hazardous gases. The voltage and current level are kept below a certain point given in table.

These voltage and current levels were determined by empirical studies of the voltage and current which causes ignition. From that ignition point, some safety factors are taken and voltage and current levels are represented in form of curve and table. Wheeler and Thornton prepared tables not only for voltage and current level but also for current to inductance level and voltage to capacitance level. The method of using the table is explained with example in Sect. 2.3.5a.

The selection of the electronic components and their operating conditions are important in order to avoid hot components. The experimentally quantified value of power dissipation is given in IEC60079-11:2011. The temperature rise for intrinsically safe circuit components is far below 1350 °C (T4) when the dissipated power is below 1.3 W at ambient temperature of 400 °C.

2.3.4 *Level of Protection of I.S. Apparatus*

As per IEC60079-11 standard, I.S. apparatus is classified into three categories:

(i) Ex 'ia': Apparatus that can bear two countable faults and still fulfills the energy limit criteria in normal operation.
(ii) Ex 'ib': Apparatus that can bear one countable faults and still fulfills the energy that criteria in normal operation.
(iii) Ex 'ic': Apparatus that is not capable of causing ignition in normal operation.

Protection level	Application zone
Ex ia	Zone 0, zone 1, zone 2
Ex ib	Zone 1, zone 2
Ex ic	Zone 2

2.3.5 *Assessment for Compliance as Per IEC60079-11*

The whole idea of intrinsic safety is to limit the spark energy and the temperature rise of the components and the other part of the circuit. To fulfill the requirements of IEC60079-11:2011, the circuit must have the following two compliances.

- Spark ignition compliance
- Thermal ignition compliance

- **Spark ignition compliance**

  Minimum Spark test apparatus Energy for different apparatus groups, above these energy level apparatuses is likely to cause the explosion in test chamber filled with explosive gases as stated by the standard.

Group I	525 μJ
Group IIA	320 μJ
Group IIB	160 μJ
Group IIC	40 μJ

The table provided in the standard is based on experimental data after performing multiple spark ignition test by Whiller and Thornton [13]. These table provide the current level at particular voltage level, inductors at particular current level, capacitors value at particular voltage level. For entire tabular data, refer Table 2.9 in IEC60079-11 standard (Table 2.10).

Table 2.9 Permitted short circuit current corresponding to the voltage and the equipment group

Voltage (V)	Permitted short circuit current mA							
	For gas group IIC apparatus		For Group IIB apparatus		For Group IIA apparatus		For Group I apparatus	
	With a factor of safety of		With a factor of safety of		With a factor of safety of		With a factor of safety of	
	×1	×1.5	×1	×1.5	×1	×1.5	×1	×1.5
12.1	5000	3330						
13.5	2320	1550	5000	3300				
14.8	1430	950	3470	2310	5000	3300		
15	1350	900	3290	2190	4730	3150		
16	1030	687	2550	1700	3630	2420	5000	3330
18	660	440	1660	1110	2240	1490	3170	2110
20	464	309	1177	785	1572	1048	2274	1516
24	261	174	650	433	896	597	1057	701
30	152	101	379	253	510	340	710	373.3
45	68	45.3	169	113	227	151	286	190.7

Table 2.10 Permitted capacitance corresponding to the voltage and the equipment group

Voltage (V)	Permitted capacitance μF							
	Gas group IIC		Group IIB		Group IIA		Group I	
	Factor of safety		Factor of safety		Factor of safety		Factor of safety	
	×1	×1.5	×1	×1.5	×1	×1.5	×1	×1.5
5.0		100						
5.6	1000	54						
6.0	600	40		1000				
8.7	47	5.9		50		1000		
9.0	40	4.9	1000	40		500		
9.5	30	3.7	700	27		255		1000
13.0	6.0	1.0	52	6.2	1000	22.5		28.5
14.2	3.8	0.70	28	4.39	270	16.4	1000	20
15.0	3.0	0.58	20.2	3.55	100	14.0	180	17.8
18.0	1.41	0.309	9.0	1.78	36	7.6	38	10
20.0	0.90	0.220	5.6	1.41	20.0	5.5	26	8
24.0	0.46	0.125	2.75	0.93	11.0	3.35	15.2	5.25
30.0	0.220	0.066	1.41	0.560	5.50	1.82	8	3.05
45.0	0.067	0.0259	0.560	0.240	1.82	0.730	3.05	1.45

Steps Involved in Assessment

Assessment of a simple inductive circuit:

Given Parameters: 20 V, Current limiting resistor = 300-Ω, Load = 1100 Ω and 100 mH.

- Power supply:
  - Step 1—Quoted minimum current limiting resistor (300 Ω).
  - Step 2—Maximum voltage of the battery or power supply circuit (say 22 V).
  - Step 3—Calculation of maximum short circuit current (22/300 = 73.3 mA).
  - Step 4—Apply a safety factor of 1.5 (73.3 × 1.5 = 110 mA).
  - Step 5—Compare in table (for IIC maximum allowable current at 22 V is 337 mA).
  - Therefore, it can be said that the assessed circuit is intrinsically safe
- Connection of loads:
  - Step 1—Total resistance after load connection (300 + 1100 = 1400).
  - Step 2—Maximum possible current (22/1400 = 15.7 mA).
  - Step 3—Apply Safety Factor 1.5 (1.5 × 15.7 = 23.6 mA).
  - Step 4—Check the inductive load table (for IIC, 100mH inductor, current for the source of 24 V is 28 mA).
  - Therefore, it can be said that the assessed circuit is intrinsically safe.

Similar assessments are done for capacitive loads as well. Spark ignition tests were performed on different points of circuit with the use of spark test apparatus to support the assessment provided in IEC60079-11. Compliance with Spark Ignition is achieved by limiting voltage, current, fault energy, stored energy, and external interference.

Reasons for spark	Protection method
High voltage across external circuit	Voltage regulators like zener barrier, regulator ICs 7805 etc.
High current in the circuit	Current limiting resistor, fuse
High fault and stored energy	Proper selection of components such as inductors, capacitors, etc.

- **Thermal ignition compliance**

  The circuits exposed to explosive atmospheres undergo testing and assessment for maximum temperatures in order to avoid triggers due to high temperatures.

  - **Assessment of Small Components**

    Temperature for small components having surface area more than or equal to 20 mm^2 may be classified as T4 (below 135 °C) if the power dissipation is below the level given in Table 2b of IS/IEC 60079-11:2006.

Assessment of wiring within apparatus

The formula used for the calculation of maximum permissible current due to self-heating is

$$I = I_f\sqrt{t(1 + aT)/T(1 + at)}$$

I	Max. permissible current
I_f	Current at which wire starts melting
t	Temperature of the wire (due to self-heating)
T	Temperature at which wire melts
a	Temperature coefficient of the wire material
a_{copper}	$0.004284\ K^{-1}$.

Alternatively, the circuit wiring can also be used on the basis of Table 3 provided in the IS/IEC 60079-11:2006. These assessments need not be performed for every wiring within the circuit, but they should be done for the wiring of supply, earth, etc. This is important because in these wirings sometimes the current levels are higher even during normal operation.

– **Assessment of PCB tracks**:

In IEC60079-11, a PCB track assessment table is provided for copper-made tracks which are highly popular in PCB making. This table applies to PCBs with a thickness of more than 1.6 mm and a single layer of a copper track with a thickness of 33 m.

After selecting the current requirement corresponding to the given thickness, these following factors are considered as per the standard to make the circuit safer.

1	Boards having thickness (0.5–1.6 mm)	Specified maximum current divided by 1.2
2	Boards having both sides conductive track	Maximum current divided by 1.5
3	Multilayer boards	Maximum current divided by 2
4	Copper thickness 18 μm	Maximum current divided by 1.5
5	Copper thickness 70 μm	Maximum current multiplied by 1.3
6	For tracks passing under components dissipating 0.25 W or more either normally or under fault conditions	Maximum current divided by 1.5
7	At terminations of components dissipating 0.25 W or more either normally or under fault conditions, and for 1.00 mm along the conductor	Either multiply the track width by 3 or divide the maximum current specified by 2

(continued)

(continued)

8	Ambient temperature below 60 °C	Maximum current divided by 1.2
9	Ambient temperature below 80 °C	Maximum current divided by 1.3

Example: If the following specifications are known PCB thickness 0.5 mm (declaration from the manufactures of PCB) Track thickness = 33 μm, No. of layer = 2, T-Class = T4, Group II and track with minimum = 0.3 mm.

Steps for calculation:

- 2.4/1.2 = 2 A (board thickness in between 0.5 and 1.6 mm)
- 2/1.5 = 20/15 = 4/3 = 1.33 A (two layer)
- 1.33/1.5 = 8/9 (copper thickness above 18 μm)
- 8/9 × 2 = 8/18 = 0.444 (termination)
- Allowable current = 0.444 A

By applying the above factors in the permissible current for width 0.3 mm.

2.3.6 Construction Requirement of I.S. Apparatus as Per IEC60079-11

a. **Enclosure requirements**: In Group I, IP54 degree of protection is required as per IEC60529. Apparatus which meets the requirement of separation as per Table 2.11 shall have minimum degree of protection of IP20.

b. **External circuit connections**: There is a limitation of L/R ratio of the external circuits which can be connected to the intrinsically safe circuit. This L_0/R_0 is determined by the following formula

$$\frac{L_0}{R_0} = \frac{8eR_s + \left(64e^2R_s^2 - 72U_o^2eL_s\right)x^{0.5}}{4.5U_o^2}\ \mathrm{H}/\Omega$$

where

R_s = Power source minimum output resistance (Ω)
U_o = Open circuit voltage (max.) (V)
L_s = Power source maximum output inductance (H)
C_s = Power source maximum output inductance (F)
If $L_s = 0$ which is quite common in the circuit
Then

$$\frac{L_0}{R_0} = \frac{32eR_s}{9U_o^2}\ \mathrm{H}/\Omega$$

Table 2.11 Maximum permissible current corresponding minimum track width

Minimum track width (mm)	Maximum permissible current for temperature classification		
	T1–T4 and Group I	T5	T6
	A	A	A
0.075	0.8	0.6	0.5
0.1	1.0	0.8	0.7
0.125	1.2	1.0	0.8
0.15	1.4	1.1	1.0
0.2	1.8	1.4	1.2
0.3	2.4	1.9	1.7
0.4	3.0	2.4	2.1
0.5	3.5	2.8	2.5
0.7	4.6	3.5	3.2
1.0	5.9	4.8	4.1
1.5	8.0	6.4	5.6
2.0	9.9	7.9	6.9
2.5	11.6	9.3	8.1
3.0	13.3	10.7	9.3
4.0	16.4	13.2	11.4
5.0	19.3	15.5	13.5
6.0	22.0	17.7	15.4

Note The value given for maximum permissible current, in amperes, is the r.m.s. a.c. or d.c. value

a. Separation distances:
   Creepage and clearance distance: Creepage is the shortest distance over a solid surface clearance is the shortcut distance in air between two conductors (Table 2.12).

Terminals are also an important part circuit component which is used for the external and internal connections. These terminals sometimes bear high voltages; therefore, the separation distance between two terminals becomes important to avoid high voltage sparks (Fig. 2.26).

2.3.7 Components on Which I.S. Depends

a. **Rating of the components**:
   Other than devices like transformer, fuses, relays, optocouplers, thermal trips, etc. which have to operate at rating for their normal operations all the devices must operate as below.

Table 2.12 Clearance, creepage distances, and separations (IEC60079-11:2011)

1	2		3		4		5		6		7	
Voltage (peak value)	Clearance		Separation distance through casting compound		Separation distance through solid insulation		Creepage distance		Distance under coating		Comparative tracking index (CTI)	
Level of protection	ia, ib	ic	ia, ib	ic	ia, ib	ic	ia, ib	ic	ia, ib	ic	ia	ib, ic
10	1.5	0.4	0.5	0.2	0.5	0.2	1.5	1.0	0.5	0.3	–	–
30	2.0	0.8	0.7	0.2	0.5	0.2	2.0	1.3	0.7	0.3	100	100
60	3.0	0.8	1.0	0.3	0.5	0.3	3.0	1.9	1.0	0.6	100	100
90	4.0	0.8	1.3	0.3	0.7	0.3	4.0	2.1	1.3	0.6	100	100
190	5.0	1.5	1.7	0.6	0.8	0.6	8.0	2.5	2.6	0.6	175	175
375	6.0	2.5	2.0	0.6	1.0	0.6	10.0	4.0	3.3	1.1	175	175
550	7.0	4.0	2.4	0.8	1.2	0.8	15.0	6.3	5.0	1.7	275	175
750	8.0	5.0	2.7	0.9	1.4	0.9	18.0	10.0	6.0	2.4	275	175
1000	10.0	7.0	3.3	1.1	1.7	1.1	25.0	12.5	8.3	2.9	275	175
1300	14.0	8.0	4.6	1.7	2.3	1.7	36.0	13.0	12.0	4.0	275	175
1575	16.0	10.0	5.3	*	2.7	*	49.0	15.0	16.3	5.8	275	175
3.3 k	*	18.0	9.0	*	4.5	*	*	32.0	*	*	275	175
4.7 k	*	22.0	12.0	*	6.0	*	*	50.0	*	*	*	*
9.5 k	*	45.0	20.0	*	10.0	*	*	100.0	*	*	*	*
15.6 k	*	70.0	33.0	*	16.5	*	*	150.0	*	*	*	*

*No values for all voltages are proposed at present

Note Evidence of compliance with the CTI requirements of insulting materials must be provided by the manufacturer. At voltages up to 10 V, the CTI of insulting materials is not required to be specified

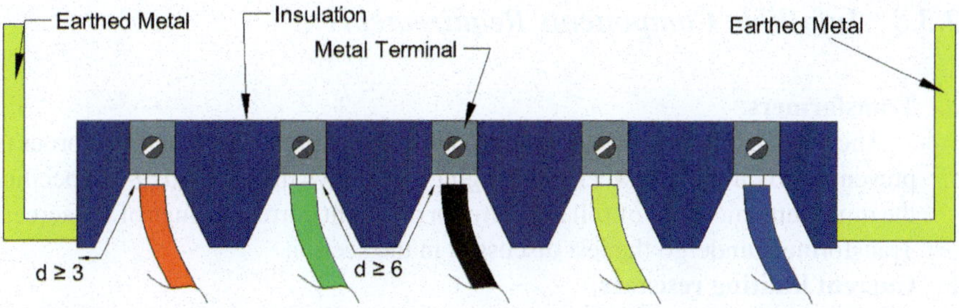

Fig. 2.26 Terminal requirement as per IEC60079-11:2011

ia and ib	ic
$I_{operation} < 2/3\ I_{max}$ $V_{operation} < 2/3\ V_{max}$ $P_{operation} < 2/3\ P_{max}$	$I_{operation} < I_{max}$ $V_{operation} < V_{max}$ $P_{operation} < 2/3\ P_{max}$

The ratings are allowed to be higher for the purpose of charging, data, downloading/uploading but only in non-hazardous area.

b. **Fuses**:
In most of the intrinsically safe circuits, fuses are present to protect the other components of the circuit by limiting the current.

- Fuses shall be encapsulated.
- $I_{braking} > (I_{operation})_{max}$ for proper operation.
- $(I_{rated})_{fire\ braking} \geq I_{operation} \times 1.5 \times 1.7$.
  (U_{rated}) fuse $\geq U_m$ or U_i
  $(P_{rae}) \geq 1.5 \times (1.7 \times I_n)^2 \times R_{CLR}$

c. **Batteries**: Batteries permitted batteries as per IEC60079 these batteries are sealed on the basis of their performance like electrolytic leakage and ventilation, voltage, internal resistance, etc. housing.

d. **Piezoelectric devices**
Since piezoelectric devices are active in nature and generates power and stores energy in crystal in form of capacitor. Many of the today's sensors comes up with piezoelectric crystal. The maximum stored energy at maximum measured voltage is given as

Group I	1500 μJ
Group IIA	950 μJ
Group IIB	250 μJ
Group IIC	50 μJ

There must be a provision for the protection of crystal from high impact which can create more energy than the above.

2.3.8 Infallible Components Requirements

a. **Transformers**:
The winding seals of transformers follow the separation distance. In order to prevent short circuits between the winging and screw, the IEC60079-11 specifies the minimum thickness of foil and wire for different current ratings of the screen. Transformers undergo the test discussed in the section.

b. **Current limiting resistors**:
Current limiting resistor circuits the incoming currents that make the circuit safer and fuse resistances should be anyone of these types.

   i. Film type
   ii. Wire wound type
   iii. Printed resistor with coating

The rating of these resistors should be at least 1.5 times of voltage and power dissipates than in normal operation (Fig. 2.27).

c. **Blocking capacitors**:
Rating of Blocking Capacitor shall be 1.5 times that of in normal operation. Reliability is a big concern in blocking capacitors which is why these capacitors should be of solid dielectric type capacitors. Like electrolytic or tantalum shall not be used for the purpose of blocking.

d. **Safety shunts**:
Zener as a voltage regulator is highly popular as a safety shunt assembly since it maintains the output voltage and also acts as a current sink when excess current is supplied. Zener barrier must always operate in reverse bias and at operating voltage V_z. To increase safety multiple layers of the barriers are used (Fig. 2.28).

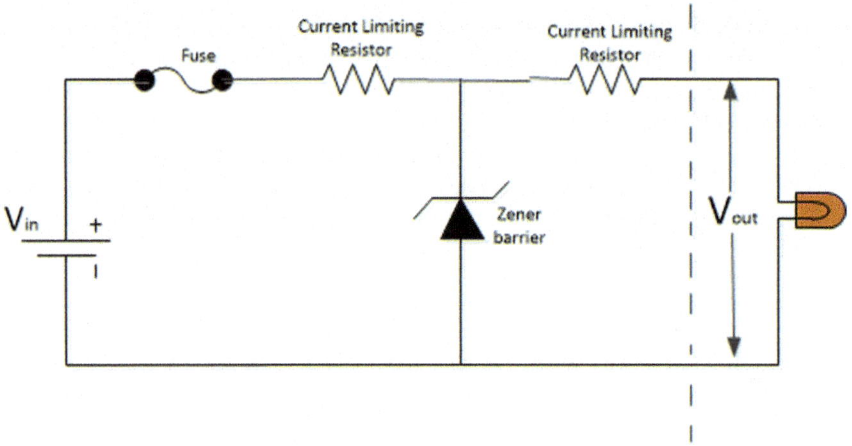

Fig. 2.27 Protection with a current limiting resistor and zener barrier

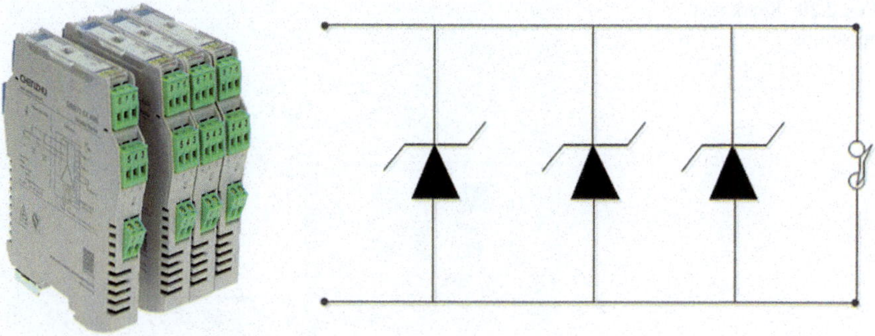

Fig. 2.28 Multiple zener barrier shunting

2.3.9 Specific Test Requirements as Per IEC60079-11

1. **Spark ignition test**
   A spark ignition test ensures that the circuit under test cannot cause ignition when exposed parts are connected to a spark test apparatus under specified conditions.
   The specified conditions for safety factor 1.5 is given in Table for the different gas groups. This table shows the required calibration current and percentage of air and oxygen mixture (Fig. 2.29).

   a. **Temperature test**
   Temperature rise test is conducted on a reference ambient temperature of 40 °C. Temperature may be measured using any suitable method.
   The temperature classification is obtained according to the max. obtained temperature for temperature classification of winding rise in temperature is calculated using the following equation.

$$t_r = \frac{R_w}{R_{wa}}(K + t_1) - (K + t_2)$$

   t_r = Temperature rise
   R_w = Max. winding resistance under test current conditions
   R_{wa} = Winding resistance at amb. Temp. t_1
   t_1 = Amb. Temp. while measuring R_{wa}
   t_2 = Amb. Temp. while measuring R_w.
   $K = \frac{1}{\propto}$, $\propto$ = temp. coefficient of resistance at 0 °C.

$$K = 234.5 \text{ K for Cu}$$

   b. **Dielectric strength test**
   This test is performed to check the voltage breakdown strength of the non-metallic dielectrics in the circuit.

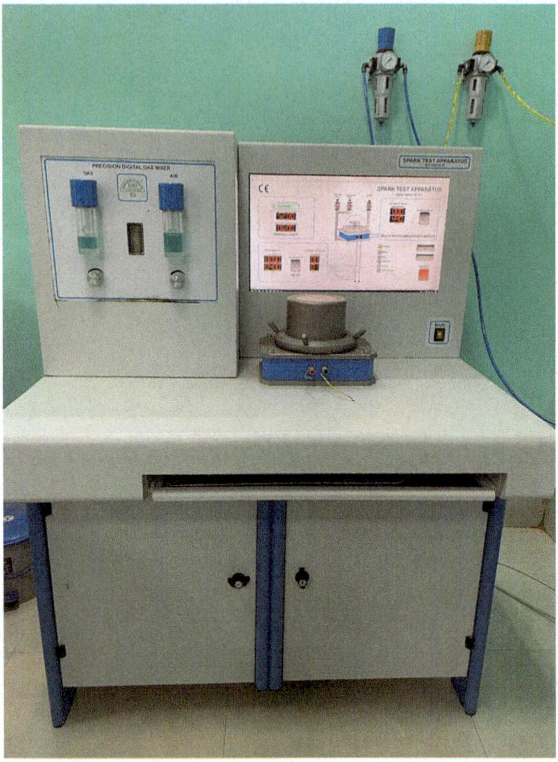

Fig. 2.29 Spark test apparatus

Test voltage may be either of anyone

1. A.C. with frequency ranges between 48 and 62 Hz.
2. D.C. with nipple below 3% (P-P) at 1.4 times the specified A.C.

While a constant V_{test} is supplied, the current flowing shall not exceed 5 mA (Fig. 2.30).

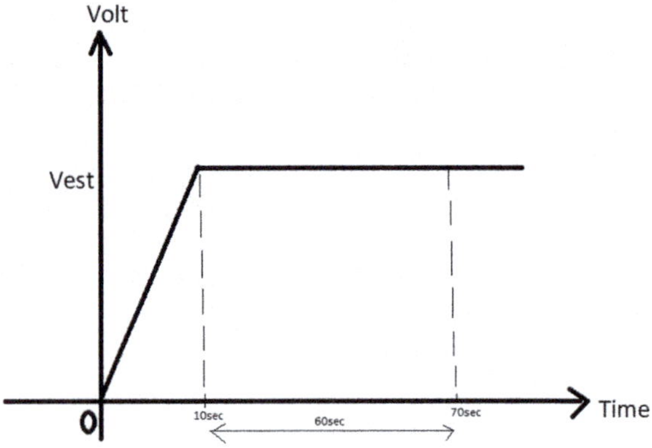

Fig. 2.30 Test voltage for dielectric strength test

2. **Mechanical tests**

   For the mechanical test, 30 N of force is applied to the enclosure using a 6 mm flat-surfaced rod for 10 s. No permanent damage should occur during this test.

3. **Cable pull test**

   This test is performed to test the strength of the cable connection with the enclosure. A cable is pulled using a force of 30 N in the direction of cable entrance for at least 1 h in this test. During this test, the acceptance criteria is that there should not be any visible displacement of cable termination but while testing the sheath may be displaced.

   Other than these above tests, there are few tests which are performed like Transformer test, Type test for diode safety barrier and test for cell and batteries.

2.3.10 Simple Apparatus

Simple apparatus is a circuit or part of a circuit that does not generate energy on its own above a certain level (1.5 V, 100 mA, and 25 mW). A simple apparatus consists mainly of passive components such as junction boxes, switches, resistors, and simple semiconductor components. A simple apparatus is exempt from the Ex marking. It is important to ensure that these appliances can withstand the voltage and current level while maintaining a separation between I.S. and non-I.S. When a simple appliance is part of a larger circuit, its entire circuit is assessed.

2.3.11 Associated Apparatus

Equipment contains both intrinsically safe and non-intrinsically safe circuits, but is designed so that the non-intrinsically safe circuit cannot adversely affect the intrinsically safe circuit. There must be a sufficient physical and galvanic separation between IS and non-IS circuits. Normally, the associated apparatus runs over a power line so it should be kept in flameproof enclosures (Fig. 2.31).

2.3.12 Marking

All the intrinsically safe apparatus shall be marked with the symbol Ex ia, Ex ib or Ex ic. Apart from Ex mark, all the relevant parameters like Uo, Li, Ci, Lo, Co shall be marked.

There must be one or more warning marks clearly visible on the body of the apparatus enclosure. An example of one such warning is as follows:

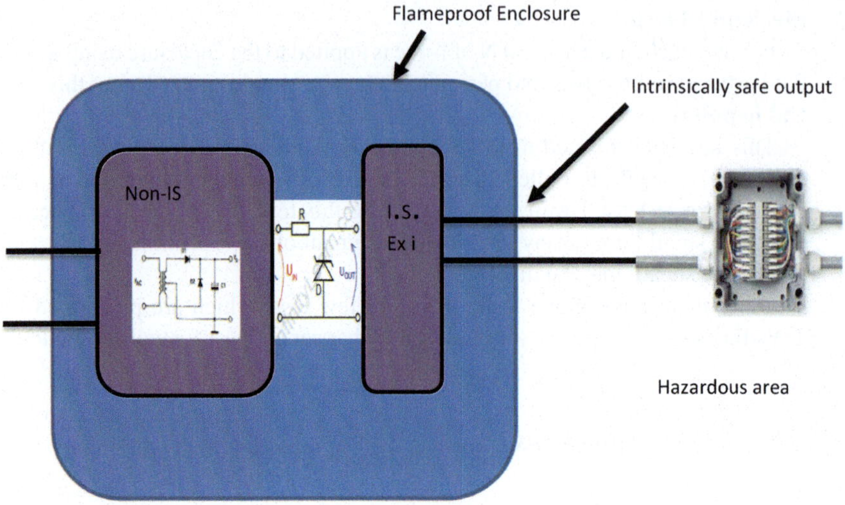

Fig. 2.31 Associated apparatus

WARNING-DONOT CHARGE THE BATTERY IN THE HAZARDOUS LOCATION

Examples of Markings [11]:

(a) **Self-contained intrinsically safe apparatus**

XYZ. LTD
PAGING RECEIVER TYPE 3
Ex ia IIB T4
$-25\ °C \leq Ta \leq +50\ °C$
Certificate No. IECEx ExCB 04****
Series No. XXX

(b) **Intrinsically safe apparatus designed to be connected to other apparatus**

XYZ. PVT.LTD
TRANSDUCTEUR TYPE 12
Ex ib IIC T4
Cert No.: Ex21xxx
L_I : 15 μH C_i : 1000 pf
U_I : 24 V I_i : 200 mA
P_I : 1.3 W

(c) **Associated apparatus**

XYZ. PVT. LTD
STROMVERSORGUNG TYPE 4
[Ex Ib] I
Cert No.: Ex09xxx
U_M 240 V
I_0 : 100 mA P_0 : 10 W
L_0 : 20 mH U_0 : 24 V
C_0 : 4.7 μF
Serial No. XXXX

(d) **Associated apparatus protected by a flameproof enclosure**

XYZ. PVT. LTD
Ex d [ia] IIB T6
ACB No: Ex05****
U_m: 240 V
U_0 : 36 V P_0 : 0.9 W
C_0 : 0.30 μF I_0 : 100 mA
L_0 : 15 mH
Serial No. XXXX

2.4 Increased Safety Ex 'e'

2.4.1 Introduction

The product reduces explosion risk in hazardous areas by eliminating arcs, sparks, and hot surfaces. Concept of increased safety was developed for the first time in Germany since 1970 [14]. UK accepted this protection concept and formulated a national standard in 1973, BS 4683 [15], Part-4.

As a result of proper design of increased safety products, the generation of arcs, sparks, and hot surfaces is prohibited. In order to avoid arcs, sparks, static charges, etc., the creepage, clearance, and insulation of the increased safety product are maintained as per code requirements. The mechanical integrity of these products is not as strong as flameproof products. During temperature classification, the maximum temperature of the component mounted inside the electrical equipment is considered. It is recommended that these products have an ingress protection rating of at least IP 54. Increased safety products do not use fuse or switchgear, such as arcing devices. By using non-sparking components, ignition may be prevented. For manufacturing increased safety products, materials such as fire reinforced polymer (FRP) and glass reinforced polymer (GRP) that do not hold static charge are preferred. Electrical equipment like terminal housings, luminaries, and motors comes under this type of protection. This type of protection was also developed for high-power equipment. IEC60079-7:2015 requires the use of good quality materials with well-defined insulation properties.

While IEC code IEC60079-7:2015 permits the use of increased safety equipment in gas group II zone 1 locations, the Bureau of Indian Standards (BIS) does not allow it and restricts its use to gas group II zone 2 only. This stipulation covers major Ex equipment like luminaries, motors and terminal housing. The above stipulation leads to additional costs to users due to the use of Ex d equipment in place of Ex e equipment in zone 1.

2.4.2 Limiting Temperature

It is the maximum permissible temperature for electrical equipment or part of electrical equipment and is lower by:

- Auto-ignition temperature of the flammable media.
- Thermal stability of the materials used.

2.4.3 Ratio Starting Current

I_a/I_n this is the ratio of initial starting current I_a and rated current I_n. In any case, the value of I_a/I_n should not be more than 10.

2.4.4 Safe Time t_E

Time t_E represents the moment during starting or stalled conditions when either the stator or rotor temperature rises to a critical level. The value of t_E should not be less than 5 s.

2.4.5 General Requirements for Ex 'e' Equipment

Increased safety enclosure Ex 'e' does not need to withstand explosion pressure as flameproof products do. A minimum IP 54 rating is required for Ex 'e' type products. There is a rubber gasket to prevent dust and liquids from entering. The ex 'e' product can be constructed from non-metal plastics with anti-static properties and high mechanical strength. The mechanical strength of the enclosure should be tested as per the procedure led down in IEC60079-0:2017.

2.4.6 Drop Test

An enclosure made of plastic needs to pass a drop test from a height of 1 m over a concrete floor. There should be no damage or deformation.

2.4.7 Terminals and Wiring

The terminal and wiring should be vibration-proof and clearance/creepage should be met as per IEC60079-7: 2015 [16].

2.4.8 Temperature Class

The temperature class of Ex ‘e’ equipment should be graded from T1 to T6 for increased safety.

2.4.9 Cable and Cable Entry

The cable should be properly anchored, plastic cable glands can be used for cable entry and used enclosure should pass IP54.

2.4.10 Increased Safety Lightings

A lamp cap and lamp holder combination of Ex ‘e’ lighting should form a flameproof joint that conforms with IEC60079-1:2014, commercial type of lamp cap and lamp holder combinations may not comply.

2.4.11 Increased Safety Fluorescent Fitting

The only fluorescent tube to be used is a cold cathode fluorescent tube, i.e., the single pin type. Ordinary starters, even if hermetically sealed, are not permitted. Therefore, only starter-less temperatures should be used. This is the reason electronic blast is only used. They can operate without a starter. In accordance with IEC60079-1:2014, the lamp holder and the pin of the fluorescent tube should form a flameproof cylindrical joint (Fig. 2.32).

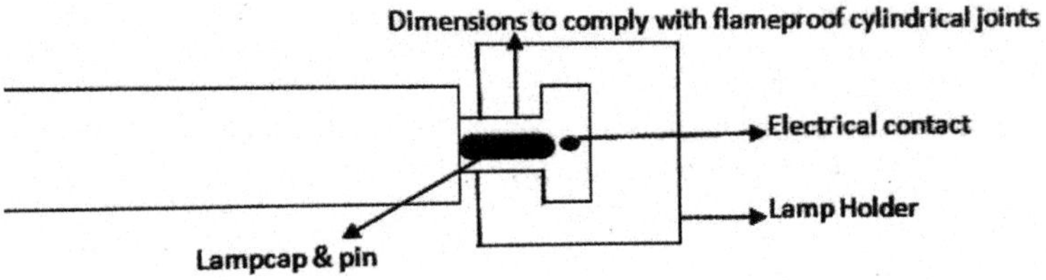

Fig. 2.32 Increased safety tubelight

2.4.12 Requirement for Increased Safety Luminaries

According to IEC60079-0:2017 and IEC60079-7:2015, the following requirements should be met by increased safety luminaries.

A. Material of construction as per code requirement.
B. Enclosure of increased safety luminaries should be IP54. However, IP65 and IP66 fittings are also available in the market.
C. The enclosed break holder shall be used.
D. Glass luminaries should pass the thermal shock test and impact test.
E. The plastic used as a light-transmitting part in luminaries shall withstand hydro testing, thermal endurance to heat and cold, and impact testing.
F. An anti-loosening and vibration-proof terminal should be used.
G. Light-transmitting parts must be covered with a 50 mm × 50 mm guard. If the mess size exceeds 50 mm × 0 mm, the fitting will be treated as unguarded.
H. The fitting shall withstand the impact test.
I. Cable entry shall be through a certified increased safety cable gland.
J. The fitting as a whole shall be type tested and certified by the testing house for increased safety including assigning a T-rating.

Due to the high lamp temperatures (T-class) of HID lamps, luminaires classified for temperature class T2 are not suitable for hazardous areas such as oil and gas industries but can be used for hydrogen panels, where hydrogen is the only combustible gas.

2.4.13 Increased Safety Motor

In gas processing plants, induction motors with cage rotors are employed. These motors do not produce arc, spark and hot surfaces during starting or running. During starting or stalled conditions, the temperature rise of either the starter or the rotor becomes critical. This is known as the time t_E. In no case should the time t_E be less than 5 s, nor should the starting current ratio I_a/I_n be greater than 10.

The limiting temperature of stator winding totally depends upon the insulation grade of the stator. As per code, 1850 °C is the maximum temperature for E class

Table 2.13 Value of safe time t_E with respect to I_A/I_N

I_A/I_N	Minimum value of t_E second
3	20
4	12
6	6
7	5

insulation. Limiting the temperature of the rotor depends on the temperature classification of the motor. If the motor certified under T3 class, it means temperature becomes 2000 °C. For those machines, both measured and calculated values of time t_E are acceptable. Also important is the starting current ratio I_A/I_N. Table 2.13 shows the minimum value of t_E in relation to I_a/I_n.

Both the stator and rotor time t_E should be evaluated, and the lower of the two times should be taken as the motor time t_E. The necessary protection circuit is to be installed so that power is tripped before t_E is reached.

2.4.14 Battery Box for Moving Vehicle

Battery-operated forklift with traction motor, pump, and lighting. The enclosure of the battery with a lower grade of protection is acceptable as there are proper ventilation openings for purging hydrogen gas. The rate of injection of hydrogen gas inside the battery continues to comply of increased safety:

$$H_2 \text{ rate} = \text{Cell capacity(AH)} * 5 \text{ cm}^3 \text{ per h}$$

The main purpose is to ensure that the H_2 concentration inside the battery enclosure should not exceed 2%. Sparks must not occur when cells and batteries are connected. Total cell connector to be rated to carry the designed current. To connect a battery pack to a plug and socket, the plug must not be removed from the socket in a hazardous area. The battery should not be recharged in a hazardous area.

2.4.15 Electrical Heating Device (EHD)

These are used for maintaining the temperature of fluids flowing through pipes. They are available in two types: constant watt and self-limiting and consist of two conductors connected to heating elements. A jacket of insulation is applied over the cable constructed thus followed by a metal braiding for mechanical protection and earthing. Sometimes the external metal braiding is additionally provided with another insulation jacket for further mechanical strength and prevention of corrosion.

Types of heat tracer cable

I. Constant watt type.
II. Self-timing type or self-regulating heat tracer cable.

2.4.15.1 Constant Watt Type

The constant watt heat tracing cables are designed to reach a temperature depending upon the resistance of the sheath conductor. Throughout the operating period, the resistance of the sheath remains the same; therefore, the dissipation of power also remains the same. The main benefit of a constant watt type heat tracing cable is its simple circuitry and low cost due to the absence of a wattage regulator circuit. Sometimes the constant dissipation of power might be a waste of power if the requirement is unsuitable. Choosing a heat trace cable of constant watt type that is suitable for the application becomes difficult because the temperature is not limited. Depending on the application, self-regulating cables are used to limit the cable's temperature.

2.4.15.2 Self-regulating Heat Tracing Cable

Electrical heat tracer heating cable was introduced in the 1930s [17]. Mineral insulated resistance heating cable was started in 1950. This type of heat tracer cable is used to limit the temperature as per the application. It is used on roofs or gutters to melt ice during the winter months. The article explains the properties, uses, and mechanisms of measuring the sheath temperature using the product classification method of self-regulating cables (Fig. 2.33).

1. Copper bus wires of 1.2 mm^2 cross section.
2. Bus bar with semi-conductive heating core.
3. Jacket for providing insulation, mechanical strength for cable.
4. Coated copper braid with aluminum for continuous ground path.
5. To protect corrosion, use outer jacket of UV resistant fluoropolymer.

Properties of Self-regulating Cable
A cable whose resistance varies with temperature below the set point will have a low resistance, while above the set point it will have a high resistance. These cables have two parallel bus wires which carry current but do not produce large heat. These cables

Fig. 2.33 Self-regulating heat tracing cable

are encased in a semi-conductive polymer. This polymer allows less current to flow when heated because it contains carbon. Different types of cables are manufactured by the manufacturer by varying both the carbon content and dosage. There is an inner jacket, for separating the bus wires from the grounding braid.

An additional rubber coating is used for commercial and industrial cables. It cannot be heated more in practice. Self-regulating heat tracer cables have numerous uses, such as freeze protection to maintain the temperature in piping and snow melting on drainage. Other uses like ice protection, pond freeze protection, soil warming, etc.

Determination of Sheath Temperature
Product classification method

During testing, 1.5 m length of heat tracer cable is placed loosely coiled in a forced air circulation oven. A representative sample is also placed horizontally in the oven (Fig. 2.32). The thermocouples are used to measure the sheath temperatures, and they are placed nearly 500 mm from each end. An outer thermocouple is used to measure the oven ambient temperature. The cable is operated at a rated voltage of 110%. As part of the testing, the oven temperature was raised incrementally by 150 K from the room temperature. The oven temperature and heater sheath temperature are recorded after a sufficient period of time has passed to stabilize and reach equilibrium. Table 2.14 shows the measured data of the heat tracer cable [17].

When the test start, the oven temperature raised at 15 °K incrementally from room temperature. Sufficient time is taken at reach oven ambient and heater sheath temperature to stabilize and achieve equilibrium. At each successive level unit, the temperature differently (ΔT), oven ambient, and sheath temperatures are recorded.

Figure 2.34 shows the variable between oven temperature and sheath temperature of the tracer heater. A tangent is drawn to the curve at 5 °K temperature difference point and extended it at the x-axis (oven temperature). The point where this tangent line meets with the x-axis is called maximum sheath temperature. This is verified by test results and curves obtained.

Figure 2.35 shows variation between oven temperature and heat trace cable output (W/m). It is clear from Fig. 2.36 and Table 2.14 output (W/m) of the heat tracer cable decreases as oven temperature increases.

It is used to freeze protection of pipelines and maintain the temperature above freezing by balancing heat loss. Its resistance varies with temperature, with low resistance below the set point and high above the set point. The heat tracer cable is self-regulating so it regulates its heat output and limits the maximum sheath temperature; thus, it becomes burnout-proof.

Thus maximum sheath temperature measured by the product classification method of self-regulating 15 W heat tracer cable is 72 °C which falls under the T6 categories (i.e., less than 85 °C) and output (W/m) decreases as oven temperature increases.

Table 2.14 Measurement data of self-regulating 15 W heat tracer cable

Trace type				Self-regulating type				
Rated voltage (V)				230				
Wattage (W)				15 W @10 °C				
Total length 9 mm				1.5				
				Measured temperature in °C				
Current I (A)	Voltage (V)	Wattage (W/m)	Total wattage $V \times I$ (W)	Sensor-reading (°C)—placed on tracer heater sample at 500 mm away from opposite side of power connection	Sensor-reading (°C)—placed on tracer heater sample at 500 mm away from power connection side	Oven ambient temp. (0 °C)	Max. sheath temp. (0 °C) for sensor-1 and sensor-2 reading is	Difference between tracer heater sheath and oven ambient temp. $(I_s - I_o)$ °C
0.10	254.0	17.53	26.3	55.9	56.2	26.3	56.2	29.9
0.09	254.0	14.86	22.3	58.1	58.9	41.3	58.1	17.6
0.07	253.0	11.61	17.4	64.1	63.9	56.3	64.1	7.8
0.05	254.0	8.93	13.4	71.5	71.9	71.3	71.9	0.5

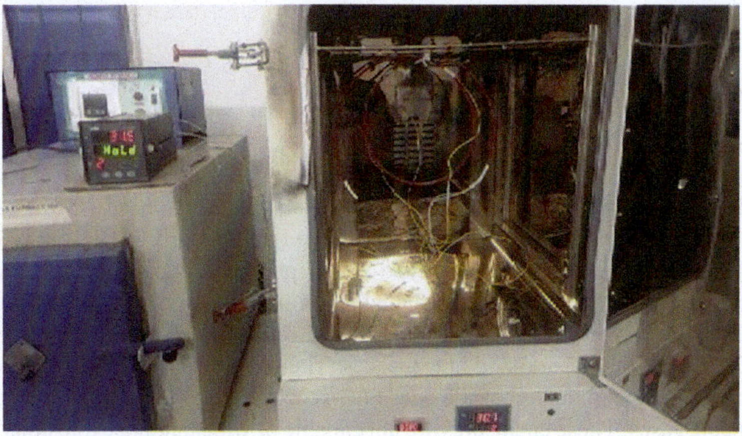

Fig. 2.34 Setup for the testing of heat tracer cable

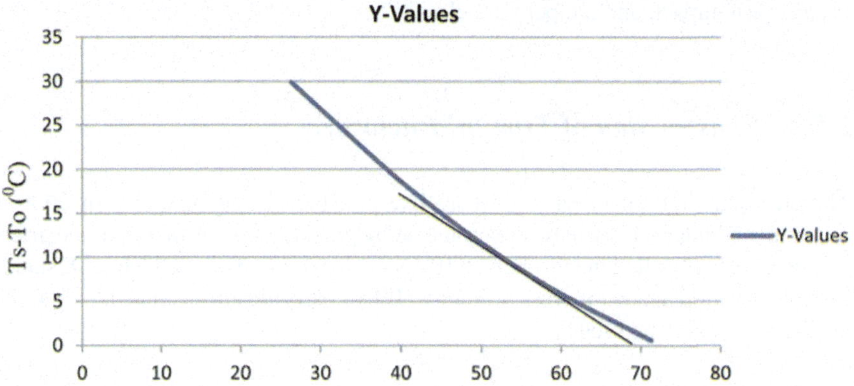

Fig. 2.35 Variation between tracer heater sheath ($T_s - T_o$) and oven ambient temperature (T_o)

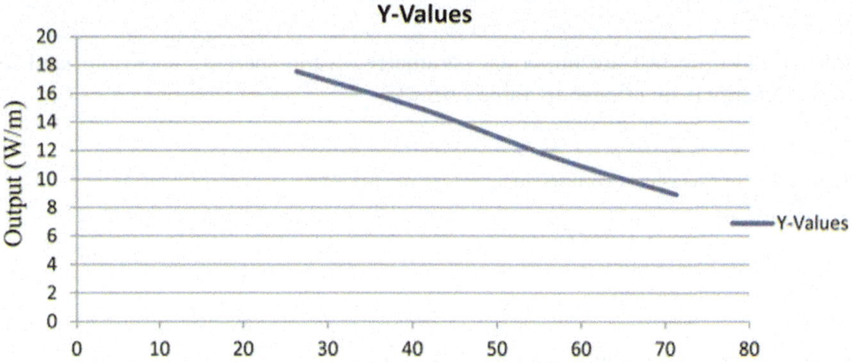

Fig. 2.36 Variation between tracer output (W/m) and oven ambient

2.5 Non-sparking

2.5.1 *Introduction*

This type of protected product is widely used in zone 2 hazardous areas. First-time term Ex 'n' was applied for the certificate issued by the test house BASSEEFA, and this certificate complies the certification standard of BASSEEFA BS 5000-3:2006 [18] for the motor. Certification of type Ex 'n' protected equipment is carried out as per IEC60079-15:2015. It is also called a non-incendive type of protection in the USA. The explosion inside the enclosure is prevented by eliminating all contributory factors like arc, spark, and hot surfaces. The surface temperature of the enclosure or component mounted inside does not exceed the auto-ignition temperature of flammable gases. These enclosures are not capable of withstanding explosions inside the enclosure. This type of protected product comprises only such components which eliminate the source of ignition. During testing, it should pass tests like IP (Ingress Protection), impact, and temperature class.

2.5.2 *Construction of Type 'n' Enclosure*

The electrical equipment having a bare live part then IP 54 type of protection is required, and internal live parts are protected by good quality insulation. If electrical equipment are installed in indoor, then IP 4X enclosures can be used where X stands for no protection against ingress of liquids. If live parts are encapsulated, then IP 20 enclosure can also be used.

2.5.3 *Mechanical Strength of the Type 'n' Enclosure*

Impact and drop test are important parameter of non-sparking enclosure, and its impact energy is mentioned in Table 2.15.

Table 2.15 Impact energy corresponding to type of enclosure

Enclosure type	Impact energy (J)	
	High risk	Low risk
Light metal or of cast metal	7	4
Plastic	7	4
Light-transmitting part without guard	4	2
Light-transmitting part with guard, to be tested without guard	2	1

Clearance and creepage distance in air between conductors should be as per code requirements.

2.5.4 Non-sparking Lightings

In non-sparking lighting, the lamp holder and lamp combination should be enclosed with a break type and tested by a testing authority, and they should be of high-quality construction. A lamp holder and lamp combination that forms an enclosed break will not have a volume greater than 20 CC. If an internal explosion takes place inside the enclosure, it should be able to withstand it. This explosion may take place due to ignition of the internal gas by make-and-break contact between holder and lamp or due to hot surfaces generated during the passing of current at contact.

If the contact mechanism between the lamp and the holder is hermetically sealed, then there will be no contact between explosive gas and spark. In this case, there will be no chance of an explosion. The sample does not need to be tested by a testing authority. Instead, it is best to use hermetically sealed make-and-break contacts in order to avoid explosion testing. Commercially available lamp holders and lamp combinations can be used in non-sparking luminaries if the enclosure is restricted breathing.

2.5.5 Restricted Breathing Type of Enclosure

As per IEC, this type of protected enclosure is used in zone 2 areas, but many countries allow it to use in zone 1 hazardous areas. This is an enclosure that is not hermetically sealed, but it is designed in such a manner that it prevents or restricts the entry of external gases.

During testing, the time interval required for an internal pressure 30 mm water gauge below the atmosphere to change to 15 mm water gauge below the atmosphere is not less than 3 min.

2.5.6 Non-sparking Fluorescent Tube Light Fitting

In a restricted breathing type enclosure, fluorescent tube light fittings with two pins hot cathodes may be used. The Starter of the tube light should be hermetically sealed and of high-quality construction can be used.

2.5.7 Temperature Class

The normal temperature of the internal or external surface of the light fitting should not exceed the temperature class defined by the testing authority at ambient 400 °C.

In the case of a restricted breathing enclosure, the internal temperature of components like the lamp is not required. It also applies to enclosed break-lamp holders with the make and break contacts hermetically sealed.

2.5.8 Non-sparking Rotating Machines

Non-sparking rotating electrical machines are similar to Increased Safety machines. The only differences are as follows:

- When starting, the rotor and stator can exceed their limiting temperatures. So the time t_E has no meaning.
- The maximum rated temperature class for these motors should not exceed T3 (2000 °C).

The protected product is installed in the zone 2 hazardous area, and the associated equipment is installed in the non-hazardous area. Equipment located in hazardous areas has energy-limiting features such as voltage-limiting zener diodes and other components connected to shunt and series. In this case, the associated apparatus present in the non-hazardous area does not need to be intrinsically safe since a short circuit fault in the interconnecting cable need not be considered as shown in Fig. 2.37.

Because they are low powered, this type of circuit is used for control and monitoring purposes.

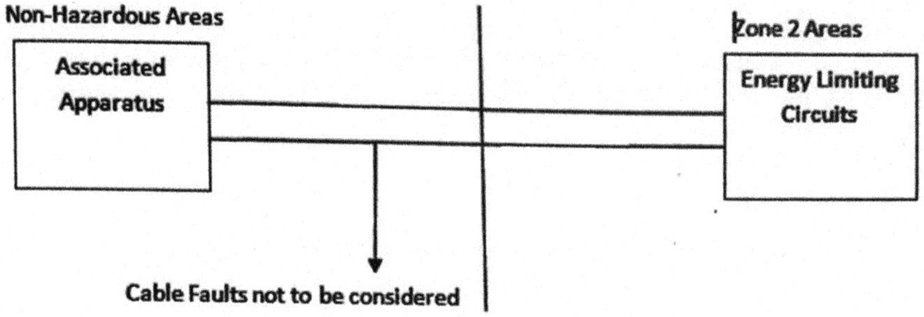

Fig. 2.37 Non-sparking Ex nA system

2.6 Purge-Protected Ex 'p'

2.6.1 Introduction

Purge-protected Ex 'p' is one of the most popular and widely used types of protection. This type of protected product prevents explosions in a hazardous area by keeping flammable media out of the enclosure. The clean air must be pumped and controlled by other equipment exposed to the hazardous area, so this protection method is expensive. The major benefit of this technique is that it is applicable for both small enclosures and complete control rooms. An enclosure can house any uncertified equipment that is purged with inert gas or fresh air. The pressure inside the enclosure remains slightly higher than the surrounding atmosphere.

This technique was developed in the UK, after the development of the oil industry in the North Sea. At that time, most drilling equipment used in the oil industry could only be protected by pressurization. This technique was referred to by BS/CP 1003, [19] Part 2 at that no specific code was formulated in the country. In the USA, this problem of the development of code was referred to the National Electrical Code (NEC) [20]. The first European Union Explosive Atmospheres Directive, 76/117/IEC [21] was published, and this problem was referred to a CENELEC standard, EN 50016 (1977) [22] which was published as BS 5501, Par 3: (1977) [23], and it provided the legal status rather than a user approach. With the second edition of the code BS/EN 50016 (1996), two new concepts were introduced. This code introduced the concept of pressurized electrical products in which flammable gases were present, and it also introduced the concept of static pressurization, which does not require ducts and gas supplies from outside, but rely only on the quality of seals. This is the protection technique used for the safe operation of electrical and electronic equipment in hazardous locations like coal mines, refineries, process industries, and defense installation, etc. where explosive vapors, gases, or dusts are present used in zone 1 and zone 2 hazardous areas.

During purging the protective gas is continuously passed through the enclosure to maintain the overpressure such that the gas mixture will never touch its LEL. This supply of protective gas should be sufficient to compensate the inevitable leakages from the pressurized enclosure and its associated ducts and tubing.

Protection employing pressurized enclosures is only as good as the associated monitoring equipment with which they are interlocked. It should be ensured that these devices are foolproof and fail-safe.

As per BS/EN 50016 (1996), there are three techniques that may ensure the pressurization of electrical equipment. In the first case, electrical equipment is pressurized outside of flammable media, and it completely depends on the seals on the electrical equipment/enclosure. By using leakage compensation, second-generation electrical equipment remains under pressure without depending on seals. During this period of time, the pressurizing medium remains connected in order to allow entry of pressurizing gas so that leakage compensation and third pressurization of electrical equipment can be sustained (Table 2.16).

Table 2.16 Type of pressurization with respect to release conditions

Release conditions	Type of pressurization					
	Static		Leakage compensation		Continuous dilution	
	Air	Inert[2]	Air	Inert[2]	Air	Inert[2]
No release of explosive gases	N[1]	Y	Y	Y	Y	Y
Deliberate release explosive gases	N[1]	N[3]	N[4]	N[4]	Y[6]	N[4]
Limited release of explosive gases in fault condition	N[1]	N[3]	N[5]	N[7]	Y[6]	Y[7]
Unlimited release of explosive gases in fault condition	N[1]	N[3]	N[5]	Y[7]	N[5]	Y[7]

2.6.2 *Condition for Pressurization*

1. As per BS/EN 50016 (1996) static pressurization with fresh air is not permitted.
2. Concentration of oxygen will be less than 1% in inert gas when electrical equipment is pressurized with inert gas.
3. Static pressure totally depends on the seals of the enclosure, so it cannot be used where there is any chance of release of flammable gases.
4. Pressurization by leakage compensation of electrical equipment with fresh air or inert gas is not applicable when release is deliberate. In such a situation, pressurization is carried out with fresh air.
5. If leakage in panel/electrical equipment is limited or unlimited, pressurization with leakage compensation by air is not permitted.
6. If the pressurization is carried out with dilution of air, the rate of airflow must be sufficient to dilute the concentration of flammable gas to less than 25% of its LEL (Lower Explosive Limit) in all.
7. After the purging of the pressurized enclosure, the concentration of oxygen in the enclosure should not exceed 2% by volume. If the upper explosive limit (UEL) of the flammable gas present in the enclosure is more than 80%, inert gas is not used for purging.

2.6.3 *Methods of Pressurization*

There are three methods of pressurization as follows:

2.6.3.1 Static Pressurization

Static pressurization involves filling the enclosure/panel with inert gas at a pressure and ensuring that it is completely sealed. In static pressurization, an inert gas is used, but its oxygen concentration should not exceed 1%. Air is not use in static pressurization. The combustion process occurs in the enclosure when there is more oxygen present. Due to an electrical fault explosion may take place inside the enclosure and allow damage to the electrical products even after the isolation of the electrical supply. During this type of pressurization, all air inside the enclosure/panel should be evacuated and the enclosure should be pressurized outside the hazardous area. Purging will be successful if oxygen concentration has fallen to less than 2% as per BS/EN 50014 (1997). After pressurization due to leakage, a pressure falls up to 50 N/m^2 inside the enclosure then the electrical supply should be switched off. Finally, it is the user’s responsibility to purge the enclosure. If an enclosure is pressurized by inert gas, an asphyxiation risk persists. Therefore, all users of this technique need to be informed.

2.6.3.2 Pressurization with Leakage Compensation

This is the most economical and popular method of pressurization. This method is more secure than static pressurization. Air is generally used as purge media so this is more economical. In the failure of static pressurization, it is necessary to ensure that the ingress of any flammable gas is sufficiently slow to permit any residual charge as per BS/EN 50014.

2.6.3.3 Pressurization with Continuous Dilution

It is an expensive method of pressurization with respect to the static and leakage compensation pressurization method. If leakage is more than continuous dilution type of mechanism of pressurization is adopted to maintain the pressure above 50 N/mth^2. Before the operation of the enclosure or electrical equipment, this continuous purging technique is applied to ensure that the interior enclosure is free from explosive atmospheres.

There are two main reasons to apply this pressurization technique, first where the leakage is great and second personnel enter into analyzer houses and rooms.

2.6.4 Overpressure and Rate of Flow of Protective Gas

As per standard IEC60079-2:2010 minimum overpressure maintained should be 5 mm Wc or 0.5 mbar. The rate of flow should be controlled so that a minimum overpressure is maintained (Fig. 2.38).

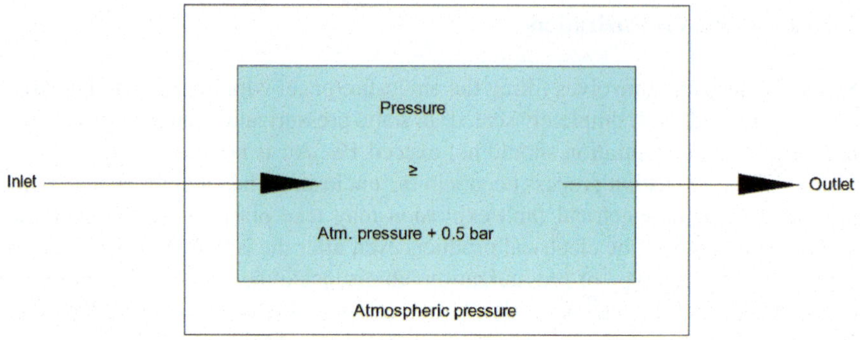

Fig. 2.38 Pressurized enclosure

2.6.5 *Construction of Enclosure and Associated Ducts*

The enclosure, ducts, and coupling components should be able to withstand the overpressure of 5 mm Wc. Flame retardant and self-extinguishing materials are used for the enclosure, duct and coupling components. These are not affected by flammable gases or the vapor in which they are used. The glass used for inspection windows should be toughened and pass a thermal and impact test. The enclosure cover and tubing should be made of MS/SS/Brass.

Cable entry in the pressurized enclosure should be through a flameproof double compression cable gland. An overpressure of 5 mm Wc should be maintained in the inlet and outlet tubes for protective gas. A special type of tool is used to open the door and cover the pressurized panel.

Ingress Protection

The minimum degree of protection with respect to dust and liquid for a pressurized Ex p enclosure should be IP40.

Cable Entries and Sealing

Proper sealing of cable entries and conduit connections are required for maintaining ingress protection and minimizing leakage.

Markings

The following information should be displayed on the appropriate places of the system:

1. Temperature class for the whole system including equipment protected by other than Ex 'p'.
2. The volume of the pressurized enclosure and associated ducts.

3. The rate of flow of the protective gas in association with the total volume and the time required to set for the time delay relay.

Warnings
To permit inspection in service, doors or covers shall bear a warning to the effect: ‘ISOLATE SUPPLY BEFORE OPENING’.

Other warnings given by the manufacturer of the Ex ‘p’ panels should be read carefully in the operation manual.

Temperature Limits
The maximum surface temperature of the enclosure should not exceed the auto-ignition temperature of the surrounding flammable media. The installed electrical equipment temperature should not exceed the temperature class defined for the pressurized panel/enclosure.

2.6.6 Graphical Presentation of Level of Overpressure

The enclosure that provides pressurization is connected to the inlet and outlet ducts. As the air compressor is connected to the inlet ducts, any leakage of hazardous gases or vapors into the system should be prevented.

The exhaust duct may be opened in the explosive zone or outside the explosive zone, depending upon the provision of a safety valve.

It is clear that the minimum overpressure maintained anywhere in the pressurized panel including ducts should be 0.5 mbar.

2.6.7 Safety Provisions

As can be seen, the security of this protection concept depends upon monitoring and control devices, which ensure proper purging and pressurization are maintained.

In these three types of pressurization, the following sequence of purging should be followed:

1. The minimum flow rate and minimum pressure should be identified by the detector before starting the purging.
2. If during purging pressure or purge flow rate falls below the value specified in the code, their control circuits should recommence purge to achieve the minimum condition.
3. After completion of the purging, the control system should close the purging exit valves and continue to monitor pressure. This is if pressurization is carried out by leakage compensate by a continuous dilution mechanism.

4. On failure of pressure or flow rate, a provision must be made by user for an alarm to be sounded or complete isolation of electrical circuit.
5. During maintenance, if the enclosure is opened then electrical equipment/ enclosure must go through its entire purge cycle on the enclosure as soon as purging is completed.

The safety devices, which ensure that the pressurized equipment cannot cause an explosion, must themselves be protected so that they do not cause explosions. Below is a list of safety devices needed to operate the Ex ‘p’ panels safely.

1. Time delay switch
2. Pressure gauges pneumatic type
3. Power cutoff switch
4. Audible/Visual indicators
5. Solenoid valves

The complete control instruction covering all such safety devices is built up to ensure the proper and safe functioning of these safety devices at the moment.

In all cases, the safety devices mounted outside the pressurized panels should be protected by any type of Ex protections suitable for the selected area.

Time Delay Relay

Time delay relay switches allow for a countdown timer to be used to ensure that the pressurized panel is completely purged before energizing the circuits in the enclosure. The time is selected such that a minimum of 5 times the volume of the protective gas to the volume of the enclosure and associated ducts and tubing is purged through the Ex p panels to ensure the concentration of the explosive gas is reduced below the L.E.L. The time delay relays are mounted inside the flameproof enclosure or control stations.

Pressure Gauges and the Switch Sockets

The pneumatic pressure gauges are meant for measuring the pressure inside the chamber, and also giving feedback to the control system. Generally, these pressure gauges are connected to FLP switch socket assemblies. For example, the overall pressure range can be divided into three levels. Very-High Pressure Level (i.e., above 0.5 mbar), Low-Pressure Level (i.e., 0.5 mbar and up to 0.25 mbar), and Very Low-Pressure Level (i.e., 0.25 mbar and below). The warning devices can be divided as per the pressure levels. Generally, hooters are used for the LPL (Low-Pressure Level). When VLPL (Very Low-Pressure Level) is reached, the power cutoff switch automatically cuts power.

Audible/Visual Indicators

Audible devices such as sound hooters are used to attract operators at a distance. Flameproof (FLP) indicating lamps are used to indicate the status of the operation.

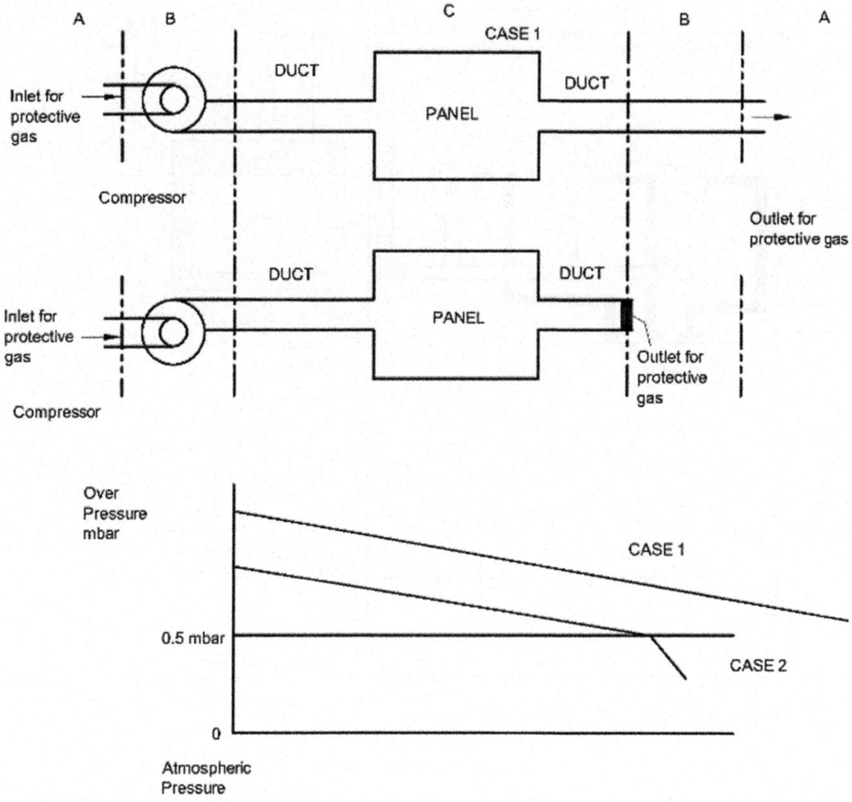

Fig. 2.39 Schematic arrangement with the overpressure level

Solenoid Valves

Flameproof solenoid valves are used to prevent the entry of explosive gas inside the Ex 'p' enclosure during exhaust of the protective gas from the exhaust ducts as in the Case 2 of Figs. 2.39 and 2.40.

2.7 Encapsulation Ex 'm'

2.7.1 Introduction

Encapsulation, also known as Ex 'm', involves the filling of void spaces around electrical circuits with a sealing material under a specific procedure. Hydrogen, acetylene, methane, propane, ethylene, etc., are explosive gases that are currently considered hazardous. The physical location of such equipment may prevent the use of normal electrical equipment. However, a specially designed type of equipment

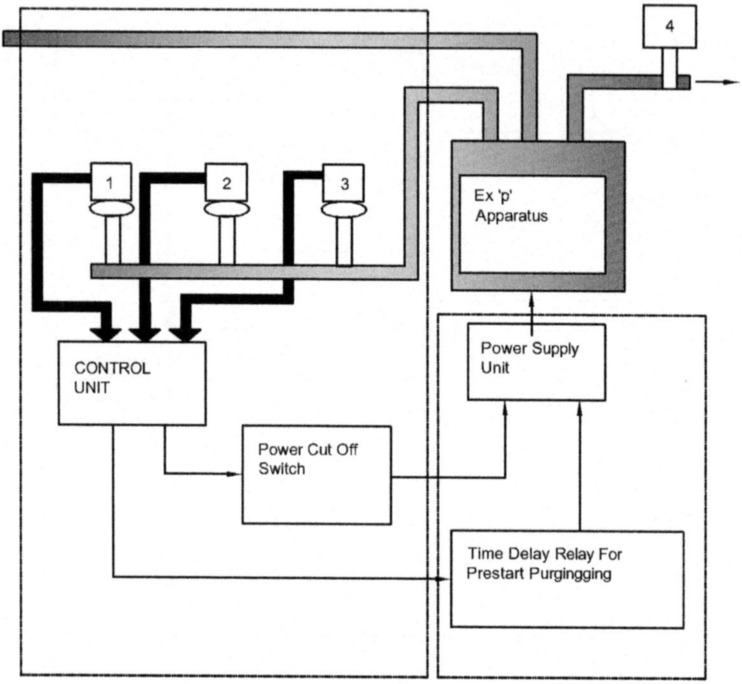

Fig. 2.40 General control unit for Ex ‘p’

called Ex ‘m’, which does not produce explosive arcs or sparks, may be used. It is a type of protection in which an explosion may be prevented by isolating the source of ignition by sealing material or setting a compound. This method of protection is applicable when the device is small or the space is constant. This technique can be applied to small equipment, solenoid valves, motor relays, PCB circuits, indicating lamps, push buttons, control switches, and electronic circuits [24]. As per IEC60079-0:2017, encapsulated Ex ‘m’ type of protected product may be used in zone 1 and zone 2 hazardous areas. In this chapter, the author discussed design and test requirements of Ex ‘m’ type of protected product.

Using potting mechanisms on all electrical parts of the equipment that may ignite flammable gases through an arc or spark, inflammable gases can be prevented from Fig. 2.41 shows how the void spaces are filled all around the electrical circuit in the schematic diagram circuit.

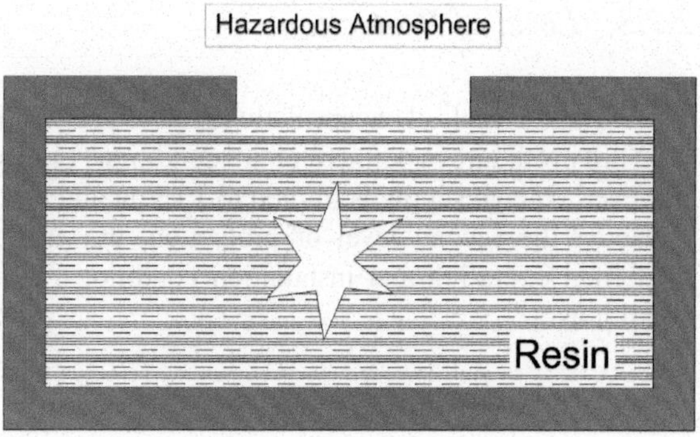

Fig. 2.41 Potting of sparking point by epoxy

2.7.2 The Encapsulation Process

There are two types of encapsulation in code. In the first type, the encapsulant is poured over the electrical circuits in a mold. Later, it is removed, leaving the encapsulant material as the outer part of the electrical products.

Second, potting in which electrical equipment is poured in the encapsulant material in liquid form for some time then it solidifies. It becomes part of the equipment and is also protected from the environment.

As per code requirement, the minimum thickness of the encapsulant between conductors and the surface of the encapsulant should be 3 mm, but if the apparatus has a surface area of less than 2 cm^2, the thickness of the encapsulant will be 1 mm. Furthermore, the thickness of encapsulant material around a void, such as a relay box, within the encapsulant must be 3 mm. If the void is less than 1 cm^3, the thickness of the encapsulant material will be 1 mm. The entire entry path of the explosive atmosphere should be sealed. As per code, 5 m of bare conductor within the encapsulant is accepted [25] (Table 2.17).

Table 2.17 Maximum distance between live part through encapsulant

Maximum voltage between live parts at maximum rated supply voltage (V)	Minimum permissible distance (mm)
380	1.0
500	1.5
660	2.0
1000	2.5
1500	4.0
3000	7.0
6000	12.0
10,000	20.0

2.7.3 Level of Protection

There are two levels of encapsulation protection. The first type ‘ma’ where protection does not invalidate in defined failure conditions or in specified abnormal conditions; however, voltage supply is limited to 1 kV at any point of the circuit.

Under the second level of protection, the encapsulation remains intact only in normal operating and installation conditions, as well as under defined failure conditions.

2.7.4 Requirements of Encapsulation Protection

An encapsulation protection system encloses the sources of ignition (i.e., electrical or electronic circuits capable of igniting) using a sealing compound that is thermally stable, mechanically strong, inert to chemical agents, and resistant to sunlight. A compound selected for encapsulation must have passed all relevant tests as defined by standard IEC60079-18:2004.

2.7.5 Properties of Compounds for Encapsulation Protection

The following are the mechanical, electrical, and thermal properties of the compound that need to be verified before selection as an encapsulating compound:

2.7.5.1 Mechanical Strength

Resistance to impact: The encapsulated apparatus should sustain impact energy as described in Table 2.18.

Here material of encapsulation is of the same requirements as a non-metallic or plastic part of the enclosure.

Resistance to drop: If the device or component is portable or hand-held type should sustain a drop test when dropped from a height of 1 m on concrete surface.

Table 2.18 Impact energy as per requirements of gas group

IS Material of encapsulation	Impact energy with risk of mechanical danger			
	Gr. I		Gr. II	
	Low	High	Low	High
	20	7	7	4

Displacement of the cable: To ascertain the proper adhesiveness of the compound with cable, a cable pull-out test is conducted by applying a tensile force (in newtons) that is either 20 times the value in millimeters of the diameter of the cable or 5 times the weight of the encapsulated apparatus, whichever is the larger value. The minimum tensile force is 1 N and the minimum duration is 1 h. Following the test, there should not be any visible displacement of the cable which could violate the type of protection.

2.7.5.2 Thermal Tests

Compounds should be physically declared safe without any changes in chemical properties and mechanical strength for a continuous storage temperature (CST) for a range of regular operating temperatures.

Temperature classification: The apparatus should be tested for the maximum surface temperature of the compound or of the internal components/circuits. The test is conducted at 110% of the rated supply voltage or current depending upon the test apparatus. To limit the maximum surface temperature a temperature protective device should be provided. If the level of protection is ‘ma’, two such devices should be encapsulated at an appropriate place under encapsulation; for ‘mb’, one such device should be encapsulated.

Temperature endurance to heat: The encapsulated apparatus should be able to withstand the maximum ambient temperature under controlled relative humidity for continuous four weeks (Fig. 2.42).

Thermal endurance to cold: The encapsulated apparatus should be able to sustain thermal endurance to cold for the lowest ambient temperature for four continuous weeks with controlled relative humidity.

Fig. 2.42 Arrangement for measurement of endurance to heat and cold for encapsulated product

2.7.5.3 Electrical Parameters

Surface insulation resistance: The encapsulating material should have a resistance of an order of 1 GΩ maximum so that static charge accumulation can be avoided.

Dielectric strength: The dielectric test is necessary to establish the isolation of circuits from each other and from their environment. According to the supply voltage of the apparatus, the test voltage can range from 500 V r.m.s. to 1500 V r.m.s. For example, if the supply voltage does not exceed 90 V peak, a test voltage of 500 V r.m.s. is required.

Discharge test: When short circuiting, the maximum surface temperature of the compound should not exceed the maximum specified temperature for the compound. No damage or deformation should be observed in the apparatus as soon as encapsulation protection is concerned.

2.7.5.4 Water Absorption Test

The encapsulating compound should not have water absorption of more than 1% by mass after putting samples of the compound under tap water for at least 24 h.

2.7.5.5 Specifications of the Compound

In general, the encapsulating compound should have the following minimum specifications:

- Information about the compound's manufacturer such as name and address.
- The compound used for potting/encapsulation should be fully described, including details of fillers and varnishing used for surface treatment.
- Insulating properties and their resistance to water absorption should be tested on the encapsulant material.
- For the water absorption test, the weight of dry samples is measured and then immersed in tap water Ex 'm' having a temperature of 230 °C for 24 h. In this case, reweighing the sample after it has dried should not exceed 1 percent.

2.7.5.6 Fault Examination and Clearance and Creepage Distances

Electrical faults disrupt normal circuit operation but do not compromise the type of protection. In general, due to design limitations, the type of protection 'mb' is limited to one fault and the type of protection 'ma' is limited to two faults. It is not necessary to investigate the fault if the clearance and creepage distances between bare current carrying parts are maintained in accordance with a standard.

2.7.5.7 Free Space and Compound Thickness Free Space

Normally, free space is not permitted in encapsulated equipment. However, if it is a design requirement or design limitation, the equipment should be pressure tested through the free space. Type 'ma' does not permit a free space of more than 10 cm^3. And also the minimum thickness of the compound is required to satisfy the encapsulation criteria.

2.7.5.8 Application of Encapsulation Protection

Sometimes it becomes difficult to make an electrical apparatus safe using another type of explosion protection. This is for example if a solar panel is to be placed in a hazardous area. It is difficult to maintain its specification in terms of output power. If it is kept in a flameproof enclosure or using pressurized protection or any other type of known protection, in this case, solar cell with its internal semiconductor transistors can be encapsulated with a layer of the suitable compound using embedding or potting.

In other cases as well, such as batteries, solenoid valves with sealed coils, sealed sensors and transducers, sealed lamps as explosion-proof components, encapsulation has been found to be easier to produce than using other types of protection.

2.7.5.9 Component Problems

Fuses: As per code, the fuse is required to be enclosed so as to avoid the ingress of encapsulant into the fuse cartridge. During breaking of such fuses, it is assumed that the temperature of electrical equipment is exceeded and the encapsulant material is not damaged.

Cells or batteries: **If the manufacturer recommends it, primary or secondary cells should be encapsulated**. During abnormal conditions, these batteries release gas, but it does not affect the concept of encapsulation.

When one takes a look at the design philosophy and examples of applications, one realizes that only sealing or potting the live circuit with a setting compound and thus assuming that the circuit is isolated from external explosive atmospheres is not sufficient to use the circuit safely in hazardous atmospheres. The encapsulation protection technique requires major considerations of environmental effects like moisture, temperature, adhesive properties of the compound with enclosure and cable sheath, design requirements as per standard, and finally validation of the design based on a series of tests conducted on the prototype sample.

2.8 Oil Immersion Ex ‘o’ IEC60079-6:2015

This type of protected product is one in which an explosive atmosphere is prevented from accessing the electrical circuits by submersing in oil. Submerging arcing or sparking components in oil to prevent flammable substances from coming into contact with the source of ignition [26]. The surface temperature of enclosures or components mounted inside does not exceed the auto-ignition temperature of surrounding flammable gases due to cooling by oil. For the first time in the UK, oil-filled protected products were manufactured according to IEC60079-6:2015 before Directive 76/117/IEC [27] was implemented. Generally, oil immersion transformers are used in hazardous areas to step-up and step-down voltage. In normal service, the transformer’s enclosure can reach its maximum level of protective oil. Similarly, the minimum level of protection that can be attained by the transformer enclosure’s liquid level in normal operation. Electronic circuits and additives extended to be immersed inside the oil have to no longer be ignition capital in ordinary operation. Figure 2.43 shows the schematic diagram of an oil-filled transformer.

2.8.1 Oil Container

The apparatus contains oil that should be properly sealed during manufacturing. The apparatus must have an indicator that shows the highest and lowest level of oil that is acceptable in all regular conditions. All fasteners need to be protected against accidental removal or loosening by personnel or vibration. As per BS/EN 50015 (1994) fasteners should be shrouds round boltheads, thread bonding, fitting of locking washers and condition of boltheads.

Fig. 2.43 Oil-filled transformer IEC60079-6:2004

2.8.2 Protective Fluid (Oil)

Protective oil is required to submerge electrical circuits to a depth of a least 25 mm within the oil. 25 mm is not a distance between the circuit and enclosure, but this distance is considered when determining the fault. If the separation distance is 25 mm, then the enclosure cannot be ruptured by electrical faults. As per Code IEC 296 (1982) [28] other oil, having the following properties may be used for oil-filled products.

1. As per IEC 836 (1988) [29] the fire point of the oil should be more than 3000 °C.
2. As per BS6664 Part 5 (1990) [30], the flash point of oil shall exceed 2000 °C.
3. As per ISO 3104 (1976) [31], the maximum kinematic viscosity must not exceed 100cST at 250 °C.
4. As per BS 5874: 1980 [32] the electrical breakdown strength of oil must be at least 27 kV.
5. As per BS 5737: 1979 [33], the volume resistivity must be at least 1014 Ω at 250 °C.
6. As per ISO 3015:2019, the pour point of the oil must be − 300 °C or lower [34].
7. As per IEC 588-2: 1978, the acidity of oil must not exceed 0.03 mg KOH/g [35].
8. Properties of material with this oil is in contact, should not adversely effected.

2.8.3 Design Requirement of Oil-Immersed Transformer

As per IEC60529, the oil-filled transformers have IP66 degree of protection, which means that they are completely dust-proof. Therefore, dust or humidity from the outside cannot enter the transformer. Transformer is completely sealed at the manufacturing end and has a pressure relief device. This oil-filled transformer operates at 1.1 times; there should be no water inside the enclosure gases/vapor should be able to escape easily from the enclosure, if evolved. Breathing is also provided in this oil-filled transformer. The maintenance requirements for the drying agent is provided by the manufacturer. As per IEC60529, the opening of the breathing device for non-sealed oil-filled product and opening of the pressure relief minimum IP23. For non-sealed oil-filled products, a distance may be used provided that in normal operation the dipstick is secured in its measurement position and the degree of protection with respect to ingress of dust and liquid will be IP66. Nearby another marking shall be provided for replacing the dipstick after use. An oil expansion facility is provided for non-sealed enclosures and may be furnished with a manually resettable protective time tool if there may be an internal fault in an oil-immersed transformer or oil-filled enclosure such would create an evolution of gas from the protective liquid. Construction of the system is such that the manufacturer can demonstrate that in normal service. The minimum filling level of protective liquid cannot fall beneath the minimum level. Clearance and creepage distance between conductors will be as per IEC60079-7 and the circuit complying with the safety requirement as per IEC60079-11:2006 of standard protection level ia or ib. The temperature at the free surface of the protective liquid should not exceed a value 25 K less than the stated minimum flash-point for the protective liquid used.

Essentially, an oil-immersed transformer is divided into four parts, such as the enclosure for power supply bushing connections and the bushing for the metering bushing. The third part is the rectifier chamber, which is filled with protective liquid and contains the rectifier stack. Finally, the fourth part is the oil tank (main tank), in which the transformer unit is housed. Transformers with safety equipment and devices for the connection of the high voltage bushing are protected by an oil-immersed protection system, and the connection for the low voltage side and metering circuit are protected by increased safety systems. External and internal earthing should be provided. Termination should be done with certified anti-loosening and vibration-proof terminals. All cable entry should be fitted with certified and approved flameproof/increased safety double compression cable glands.

2.8.4 Safety Requirement

The following are the safety device provisions required in Ex 'o' transformers.

2.8.4.1 Pressure Relief Valve

The enclosure of the transformer containing protective liquids should contain certified pressure relief devices. The safety feature of this device which permits the escape of any high pressure that may build up within the transformer oil tank under adverse condition.

2.8.4.2 Oil Level Gauge

An oil level gauge should be provided on the transformer main tank and on the HV chamber to measure the level of oil in the tank. Whenever possible, the oil level should be at least 25 mm above the live parts such as bushings, windings, etc.

2.8.4.3 Oil Drain Plug

Transformer enclosures should have an oil drain plug with an effective seal for draining out the oil.

2.8.4.4 Temperature Gauge

The certified temperature gauge should be provided on the transformer's enclosure which contains protective liquids. A temperature indicator continuously indicates the fluid's temperature. If the oil temperature increases to 700 °C, then the unit should be tripped automatically.

2.8.4.5 Creepage Distance and Clearance

Creepage distance and clearance between all connecting terminals should comply with IEC60079-7:2006.

2.8.4.6 Enclosure Body

The enclosure should be strong enough to sustain an impact test satisfactory as per IEC60079-6:2007.

Ingress Protection (IP)
According to IEC60529-2001, all enclosures of Ex o transformers must have a minimum of IP66 ingress protection.

The oil-immersed (Ex o) transformer is generally used in zone 2 hazardous area in India. According to IEC60079-6:2007, the protective oil should conform to the requirements. The basic design concept is very important to design an Ex o transformer for safe operation in hazardous areas. Ex transformers must be equipped with safety devices. All safety devices must be certified and approved for the specific gas groups.

2.9 Sand Filling Ex 'q'

Explosions may be prevented in flammable media by submerging the electrical circuit with quartz, powder or sand. Due to the quenching effect, ignition cannot propagate outside the filling materials. Filling materials like powder have a different function in this context. The ISO 565-1990 [36] specifies that the filling material may be quartz or solid glass particles with a size that passes through a metal sieve of 1 mm aperture, but not a metal sieve of 1.5 mm aperture. As a sealant, the leakage current should be less than 1 A when the electrode system is charged with 1000 V DC. To accomplish this, the electrode system should be immersed in powder or glass as filling material, surrounded by at least 1 cm of space.

Table 2.19 Distance between conducting and insulating parts [37]

Maximum voltage through the filling between conducting and insulating parts (Volt RMS AC/DC)	Separation (mm)
0–275	≥ 5
276–420	≥ 6
421–550	≥ 8
551–750	≥ 10
751–1000	≥ 14
1001–3000	≥ 36
3001–6000	≥ 60
6001–10,000	≥ 100

2.9.1 *The Enclosure*

As per IEC60079-0:2017, sand-filled enclosures should be mechanically strong, and no tests are required to verify their integrity. As per IEC60079-5:2015 [37], no filling material may escape from the enclosure, and no apertures greater than 0.1 mm may be created in the enclosure. The degree of protection with respect to dust and liquid should be a minimum IP54 as per BS/EN IEC60529. IP54 is also applicable if the sand-filled enclosure has a breathing device. During the type test, the enclosure should sustain 1.5 × 105 N/m^2 of pressure. The pressure test will be increased to 15 × 105 N/m^2 if the volume of the capacitor is greater than the one-eighth volume of the filling material.

2.9.2 *Electrical Components and Circuits Inside the Filling*

Despite the voltages being derived with possible overloads and fault conditions, such as circuit board failures, it is still necessary to separate the electrical components and circuits inside the filling from the enclosure by the distances shown in Tables 2.19 and 2.20. As relays are allowed within filling material, if they have their own enclosure which will withstand faults within them and its internal volume is less than 30 cc, if the internal volume is more than 30 cc, a 15 mm separation is applied when Table 2.19 is used.

2.9.3 *Type Testing*

A sand-filled product must undergo mechanical integrity tests, flammability tests of the material, electrical strength tests, and surface temperature tests.

Table 2.20 Separation of conducting parts within the filling [37]

Maximum voltage (peak Value) (V)	Minimum creepage distance (in mm)	Minimum comparative tracking index allowed	Minimum creepage distance under coating (mm)	Minimum Distance through filling material (mm)
10	1.5	N/A	0.6	1.5
30	2.0	100	0.7	1.5
60	3.0	100	1.0	1.5
90	4.0	100	1.3	2.0
190	8.0	175	2.6	2.0
375	10.0	175	3.3	3.0
550	15.0	175	5.0	3.0
750	18.0	175	6.0	5.0
1000	25.0	175	8.3	5.0
1300	36.0	175	12.0	10.0
1575	49.0	175	13.3	10.0

N/A CTI is not considered below 10 V

Pressure testing of products: Pressure testing of sand-filled products is carried out with 1.5 times atmospheric pressure for 1 min without any deformation and damage.

Flammability list of enclosure material: As per BS 2782:1992 [38], during testing a piece of enclosure put on Bunson burner, it should not damage more than 6 cm and the flame should be quenched before 60 s. Removal from the Bunson burner.

Testing for electric strength: 1000 V is applied on the electrode at 250 °C, 45–55% humidity, and 10^{-6} A leakage current will be measured.

Temperature: Temperature of enclosure should be measured in full load. Temperature class is measured at 1.7 times fused rated current.

Eight types of protection are used in the design of explosion-proof equipment for use in hazardous areas. The performance and construction features are also described in this chapter.

References

1. Vishwakarma RK, Singh AK, Ahirwal B, Kumar A, Kumar N, Sinha A (2010) Assessment of possible causes of failure of flameproof electrical equipment in under ground coal mines. Mine Technol 1(1):3–9
2. Gupta BK (2014) Type of protection. In: 2nd International seminar on "design development testing and certification of ex-equipment", DTEX-2014
3. McMillan A (1998) Protection concept of flameproof enclosure 'd'. Electrical installation in hazardous area

4. IEC60079-1:2014. Electrical apparatus for explosive gas atmosphere. Part-1: Flameproof enclosure 'd'
5. IEC60079-0:2017. Electrical apparatus for explosive gas atmosphere. Part-0: General requirements
6. ISO 468:1982. Surface roughness—parameters, their values and general rules for specifying requirements
7. IEC 60112:1979. Method of test for determining the comparative and the proof tracking indices of solid insulating materials under moist
8. Vishwakarma RK (2005) Dust ignition protection in hazardous locations. Training program for ONGC executives on flameproof equipment
9. McMillan A (1998) Protection concept in intrinsic safety. Electrical installations in hazardous areas. Butterworth-Heinemann
10. Bottrill G, Cheyne D, Vijayaraghavan G, Mackay S (2005) Protection concept 'i' principles. Practical electrical equipment and installations in hazardous areas. Newnes
11. Explosive atmosphere. Part-11: Equipment protection by intrinsic safety
12. Towle LC (1994) Intrinsic safety-the way forward. In: 1994 Fifth international conference on electrical safety in hazardous environments
13. McMillan A (1998) Apparatus with protection concept intrinsic safety 'i'. Electrical installations in hazardous area
14. McMillan A (1998) Apparatus with protection concept increased safety 'e'. Electrical installations in hazardous area
15. BS 4683-2 (1971) Electrical apparatus explosive atmospheres. Part 4 (1973) type of protection ex 'e'
16. IEC60079-7:2015. Equipment protection by increased safety 'e'
17. Singh AK, Bhukya P, Kumar N, Kumar A, Mishra RK (2017) Self regulating heat tracing cable. Electrical India, 3 Sept 2017
18. BS 5000-3:2006. Specification for rotating electrical machines with type of protection N
19. BS/CP 1003. Electrical apparatus and associated equipment for use in explosive atmosphere of gas or vapour other than mining operation vapor (1996). Methods of meeting the explosion hazard other than by the use of flameproof and intrinsically safe equipment
20. NEC70 (1990) National electrical code (USA), article 500, hazardous (classified) locations
21. 76/17/IEC (1975) Council directive on approximation of the laws of member states connecting electrical equipment for use in potentially explosive atmospheres, December
22. BS/EN 50016:2002. Electrical apparatus for explosive atmosphere
23. BC 5501 Part 3 (1977) Electrical apparatus for potentially explosive atmospheres. Part 3 (1977), pressurized apparatus 'p'
24. Vishwakarma RK, Singh AK, Ahirwal B (2009) Encapsulation—a prevalent protection technique for small devices and ex components. In: 1st International seminal on design, development, testing and certification of ex-equipment, DTEX-2009
25. IEC60079-18:2004. Explosive atmospheres. Part 18: Equipment protection by encapsulation 'm'
26. IEC60079-6:2015. Equipment protection by liquid immersion 'o'
27. 76/117/IEC council directive of 18 December 1975 on the approximation of the laws of the member states concerning electrical equipment for use in a potentially explosive atmosphere
28. IEC 296 (1982) Specification for unused material insulating oils for transformer and switchgear
29. IEC 836 (1988) Specification for silicon liquids for electrical purposes
30. BS 6664 Part 5 (1990) Flashpoint of petroleum-related products, method for determination of flash point by Pensky-Martens closed tester
31. ISO 3104 (1976) Petroleum products transparent and opaque liquids—determination of kinematic viscosity and calculation of dynamic viscosity
32. BS 5874:1980. Method for determination of the electric strength of insulating oils
33. BS 5737:1979. Method for measurement of relative permittivity dielectric dissipation factor and DC resistivity of insulating oils
34. ISO 3015:2019. Methods of test for petroleum and its products

35. IEC 588-2:1978. AsKarels for transformers and capacitors. Part 2 (1978). Test methods
36. ISO 565:1990. Test sieves—metal wire cloth, perforated metal plate and electro/formed sheet—nominal size of openings
37. IEC60079-5:2015. Electrical equipment for potential explosive atmospheres. Powder filling 'q'
38. BS 2782:1992. Method of testing plastics. Method 140A—determination of the Burning behavior of horizontal and vertical specimens in contact with

Chapter 3
Explosion-Proof Equipment and the Generation of Explosion Pressure Inside It

3.1 Introduction

Explosion-proof equipment requires enough strength in terms of thickness of body and cover, use of high-tensile bolts, or thread attachment between parts. These equipments can withstand the pressure developed during an internal explosion of an explosive mixture due to electrical arc, spark, or hot surface of internal components. It also prevents the transmission of the internal ignition to the surrounding of the enclosure. In this chapter, author discussed about the history of hazardous area, origin of concept of flameproof and intrinsically safe electrical products, mechanism of development of pressure in closed vessel, and the experimental investigation which are made to study effect of position of ignition source, presence of obstacle, material of obstacle on explosion pressure, and rate of pressure rise.

3.2 Hazardous Area—History

The first hazardous area was discovered in underground coal mines. This hazard was created due to the presence of a firedamp and coal dust cloud. As coal is extracted from underground coalmines, methane is released, and when it mixes with air, it creates an explosive atmosphere. When coal is stored for around 1000 h, it becomes completely free of methane or firedamp [1]. Doing something completely right always involves a phase of doing it incorrectly.

Similar experiences have been encountered in the past when dealing with dangerous areas created in underground coal mines. In the past, when miners left the mined coal in the shaft after completing their day's work, methane would accumulate in the mine's roof pocket. As a result, an explosive area would be formed, which required a solution. Unfortunately, the remedy suggested was worse than the malady.

Firemen crawled along the mine floor with their lighted torches and ignited explosive methane layers before the miners entered for the next shift. Despite the fact that

© The Author(s), under exclusive license to Springer Nature Singapore Pte Ltd. 2023
A. Kumar Singh, *Explosion-Proof Equipment in Hazardous Area*,
https://doi.org/10.1007/978-981-99-2516-2_3

the ignition itself would not necessarily result in any sort of fatality, it was the secondary ignition of coal dust produced by the methane explosion that proved to be the near disaster in most cases. Finding firearms was also a daunting task, and even criminals were not willing to volunteer for the job in return for the remission of their sentences. We even went as far as assigning ponies with candles on their saddles, engineered for the specific purpose, only to find out that the process was still inefficient.

With the invention of the Davy lamp by Sir Humphrey Davy, the mines surfaced, and forced ventilation was installed. This helped methane to be diluted with fresh air resulting in it being below its explosive limit.

As a result of the electrical equipment installed in the mines, there was always the possibility of explosions. In order to prevent an explosion, the enclosed motors were made explosion-proof. Later, this concept was applied to cast iron enclosed in tight-fitting joints. One of the electrical equipment, a low-voltage signaling bell introduced into mines. The signaling bells were believed to be safe due to the low operating voltage; however, two catastrophic explosions were traced in 1912 and 1913 as a result of the bells. In later studies, it was shown that signaling bell voltage circuits ignited mine gases, leading to the development of intrinsically safe circuit designs.

3.3 Origin of Concept of Flameproof and Intrinsically Safe Electrical Equipments

Historically, explosion-proof certification began in Great Britain in 1911 when the 'Coal Mines Act' was passed in order to regulate the use of electricity in coal mines. The primary purpose of this procedure is to prevent shocks, fires, and open sparks from occurring.

It is normal for sparks to occur at the contacts of switches and contactors, as well as at the slip rings of the commutators of electric motors. It is necessary to enclose these parts of the electrical circuit in a suitable manner. Although the spark can ignite a flammable atmosphere within the enclosure, the enclosure must be strong enough to withstand an explosion without any risk of damage and all joints should be tightly secured to prevent damage. The result is that the product of combustion forced through the joints would be cooled sufficiently to reduce its effectiveness in igniting any flammable mixture of gases in the external environment. This design of an enclosure that can withstand an internal explosion and prevent an internal explosion from communicating to the outside atmosphere was first documented in British Standard BS 229 [2] in 1926, as the standard for a 'flameproof enclosure'.

There is also a provision in the Coal Mines Act of 1911 that considers the possibility of sparking occurring on cables under fault conditions [3]. According to the regulations, electric circuits should be protected against faults or overloads. The purpose of this is to minimize the risk of open sparking and to prevent any fire hazards. Despite the fact that electrical faults cannot be completely prevented, it is desired that they occur infrequently. In the event that they occur, the circuit should be isolated immediately. The decision was also made that, excluding signaling and telecommunication apparatus, electricity would only be permitted in those parts of the mine where the concentration of firedamp in the general air was not greater than 1.25%. Whenever methane concentrations exceeded 1.25%, electricity was to be disconnected from all electrical equipment.

Consequently, the regulations set out requirements for the safe use of electricity in mines for power and lighting. Signaling apparatuses were excluded from most of these regulations at that time since they were considered to be non-incentives and could not cause ignition.

A series of coal mine explosions in Britain following the passage of the Coal Mines Act in 1911 led to a review of the regulations governing mine electricity use.

Investigations revealed that signaling circuits that had been considered non-incentives were actually incentives. The study found that the igniting property of a bell signaling circuit is more affected by the circuit's current and inductance than by the battery's voltage. With the progressive reduction of the inductance, the minimum current required to generate ignition increased rapidly, and the results were very consistent. There were several experiments conducted to determine whether bells could be made safe by reducing the inductance or current in the circuit or by safely dissipating the inductive energy.

BS 1259 [4] was the first British Standard for testing and certification of intrinsically safe products, which was revised in 1958 to BS 1259 [5].

In 1956, the National Electric Code (NEC) of the USA recognized this concept, and the Instrument Society of America published its first standard on intrinsic safety in 1965 [6].

In the early 1960s, CSIR-Central Institute of Mining and Fuel Research (CIMFR), Dhanbad, started certifying flameproof and intrinsically safe equipment. In those days, CSIR-CIMFR, Dhanbad, conducted testing and certification of coal mining equipment in order to ensure compliance with British Standards. The first Indian Standard IS 5780 on intrinsic safety was published in 1971 [7], and the first Indian Standard IS 2148 for testing and certification of flameproof products was published in 1968 [8]. In the early 1970s, the explosion-proof testing and certification process was diversified to include the surface hazardous area. Currently, 80% of the Ex certification work in India is performed by CSIR-CIMFR, Dhanbad, for surface hazardous industries such as petrochemicals, fertilizers, refineries, and defense installations and for underground coal and oil mines.

Regulatory requirements in India require third-party certification of hazardous area electrical equipment. However, the primary responsibility for ensuring the safe operation of certified equipment lies with the end users. Then, why is this certification provided by a third party? There is a simple answer to this question. Investigations have revealed that the occurrence of an explosion caused by Ex certified equipment has been rare. This may be due to the fact that the factors of safety involved are higher than those usually required. This is why end users need certified equipment more out of their own interest than out of fear of the law.

So far, we know that CSIR-CIMFR's testing and certification process is very straightforward. Ex test certificates may be obtained by a manufacturer or an authorized agent of the manufacturer. It should be noted that certification is essentially an exercise in gathering information to verify that all relevant clauses of the specified standard have been met.

The essential elements of this information are the manufacturer's drawing and the sample test results that provide evidence of compliance with certain testing norms incorporated in the standard. It is important to physically compare the sample with the drawing in order to ensure that the two are identical.

These drawings are utilized by CSIR-CIMFR during Ex testing, including comparison with prototypes or samples.

Furthermore, these drawings are used in the final certification of the products and are kept as a record. The drawing kept on record is also an essential tool for product surveillance and quality control testing performed by the Bureau of Indian Standards (BIS). In addition to providing a general outline of the equipment, these drawings are necessary to ensure compliance with standards. As per current practice, a copy of the drawings used for certification is signed and stamped before being given to the manufacturer. By doing so, it is ensured that the subsequent production will be in compliance with the specifications shown in the drawing.

In some cases, Ex certification is based on testing conducted at the manufacturer's facility, which is permitted by the standard as well. It is here that tests are conducted in the presence of representatives of the user industry and CSIR-CIMFR officials. Manufacturers are required to verify the traceability of their measurements to national standards through a comprehensive system of instrument calibration in addition to type testing. Test data are maintained in CSIR-CIMFR records, and a certificate of certification is issued.

As a result of the harmonization of Indian Standards with the International Electrotechnical Commission (IEC) and the implementation of the ATEX directive in European countries, the certification of Ex products has undergone significant changes. Although ATEX directives are not applicable to India, many products are imported from European Union countries, particularly in the mining industry. Therefore, it is necessary to review the current explosion-proof certification system.

Since the implementation of the ATEX directives, flameproof, intrinsically safe, and other products have been redefined for use in combustible dusty environments. As a result, the certifying authorities will have to ensure that a mining category of flameproof equipment can be used safely in coal dust environments. In this regard, another fundamental question will arise concerning the safety of the flameproof equipment currently in use when it comes to coal dust. Several studies have demonstrated that the gap in a flameproof enclosure must be increased four to six times for a coal dust explosion to occur. Similarly, the energy of a spark in an intrinsically safe circuit will have to be increased to 400 times to cause the ignition to coal dust cloud. Alternatively, it may be said that the current equipment in use is adequately safe and that the days of only flameproof equipment are over. It is also imperative that flameproof equipment meets dustproof specifications. In the first place, the maximum surface temperature of electrical equipment/apparatus must not exceed two-thirds of

Table 3.1 Details of the Maximum Experimental Safe Gap (MESG), minimum igniting current ratio (MIC), and ignition energy for the below-mentioned gas groups their respective reference gases

Gas group	Reference gas	MESG	MIC (ratio)[a]	Ignition energy (μJ)
IIC	Hydrogen	< 0.5	> 0.45	20
IIB	Ethylene	> 0.5 to < 0.9	> 0.5 to < 0.8	60
IIA	Propane	> 0.9	> 0.8	260

[a]MIC ratio is MIC for a particular gas to MIC for methane

the ignition temperature of the dust cloud in the combustible media. In general, 10 °C is marginally considered when determining the maximum operating temperature of flameproof enclosures. On the other hand, manufacturers will have to enhance their quality management systems in accordance with ISO 9000 standards. This may be the vision for the future.

3.4 Maximum Experiment Safe Gap (MESG)

In accordance with IEC 60079-20 [9], an apparatus consisting of two hemispheres with a volume of 20 cm^3 and a flange width of 25 mm is required. To determine the safe gap, the enclosure is placed inside a larger vessel and filled with the same gas mixture as the larger vessel. A mixture of gas is ignited within the enclosure by an artificial spark plug. The gap between the 25 mm flanges is just wide enough for ten ignitions within the sphere to fail to ignite the mixture outside the enclosure. This maximum gap is called Maximum Experimental Safe Gap (MESG). As shown in Table 3.1, the Maximum Experimental Safe Gap (MESG) and minimum ignition current ratio (i.e., the ratio between the minimum ignition current of a particular gas to the minimum ignition current of methane) are shown for gas groups IIA/IIB and IIC.

3.5 Mechanism of Development of Pressure in Closed Vessel

A study was conducted by Beyling and others to examine the mechanism by which explosion pressure develops inside closed vessels. As a result of this experiment, it was found that if a methane–air mixture is contained in a vessel containing an explosion, that explosion would not spread to the outside explosive atmosphere if the vessel has an aperture of a certain size [10]. The experiment also led to the realization that electrical equipment in flammable media should not be completely sealed in order to prevent the spread of fire. A number of prominent researchers, such as Statham and Wheeler, have demonstrated that flanges and gaps play an instrumental role in determining whether an explosion within an enclosure will spread to the outside. A thorough investigation of the effects of the volume of the enclosure, the position of the ignition source, the flange width, the flange gap, and turbulence was conducted during the research. Also, it was observed that the Maximum Experimental Safe Gap (MESG) depends largely on the composition of the internal and external gas mixtures (Table 3.1).

In the past, it has been observed that the explosion parameters such as the maximum explosion pressure and the maximum rate of change of pressure rise $(dp/dt)_{exp}$ in a closed vessel depend on the concentration of the gas–air mixture, the shape and size of the vessel, the position, and energy of the ignition source, the presence of obstacles which may create turbulence during the explosion [11–18]. We have previously reported the maximum explosion pressure (P_{max}) and rate of pressure rise $(dp/dt)_{exp}$ in cylindrical vessels with methane and air mixtures [17–21]. It is very important to study gas explosions in closed vessels in order to design and develop flameproof enclosures. During the experiment by earlier researchers, it was observed that [10] (Fig. 3.1):

(a) A pressure of 1.93 kg/cm^2 was developed at the end of 55 ms when the safety diaphragm was ruptured.

Fig. 3.1 Explosion development in a closed vessel

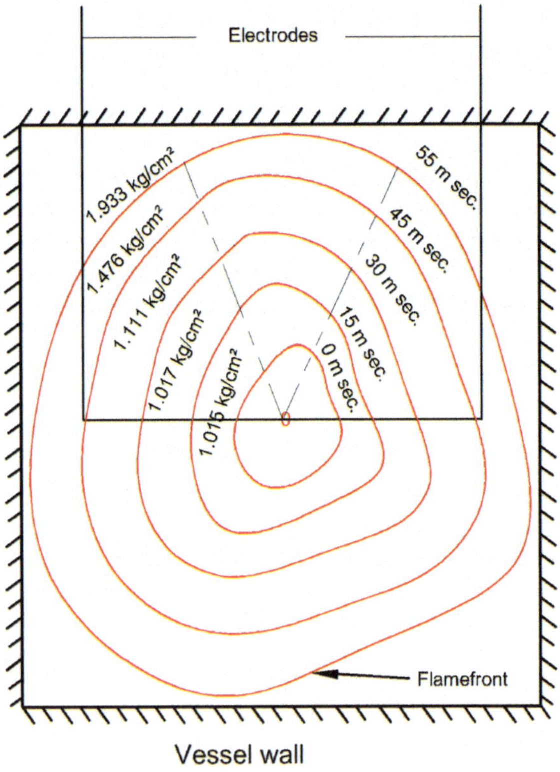

(b) A maximum pressure of 3.86 kg/cm^2 would have developed at 100 ms in a similar closed vessel if the safety diaphragm would not have ruptured.

(a) The experiment also revealed:

(c) Different sizes of enclosures are having volumes of 0.001, 1, and 20 m^3, but the maximum pressure developed in these three sizes of enclosures was 7 bar, demonstrating that the maximum pressure developed inside a sealed vessel was independent of shape and size.
(d) Figure 3.2 illustrates the rate of pressure rise $(dp/dt)_{max}$ as 720 bar/s, 75 bar/s, and 27 bar/s, respectively, for enclosures with volumes of 0.001 m^3, 1 m^3, and 20 m^3. In other words, the rate of pressure rise decreases with increasing vessel size. Figure 3.3 illustrates this concept graphically.

(e) During experiment, it was observed that the rate of pressure rise $(dp/dt)_{max}$ is directly proportional to the ratio of the surface area of the vessel to its volume.

$$\text{i.e.,} \left(\frac{dp}{dt}\right)_{max} \propto \frac{\text{Surface area}}{\text{Volume of the enclosure}}$$

The rate of pressure rise in a non-spherical vessel is greater than that in a spherical vessel of equal volume. Therefore, the product of the maximum rate of pressure rise and the cube root of the vessel volume is constant.

$$\left(\frac{dp}{dt}\right)_{max} \times V^{1/3} = \text{constant} = \text{kg}$$

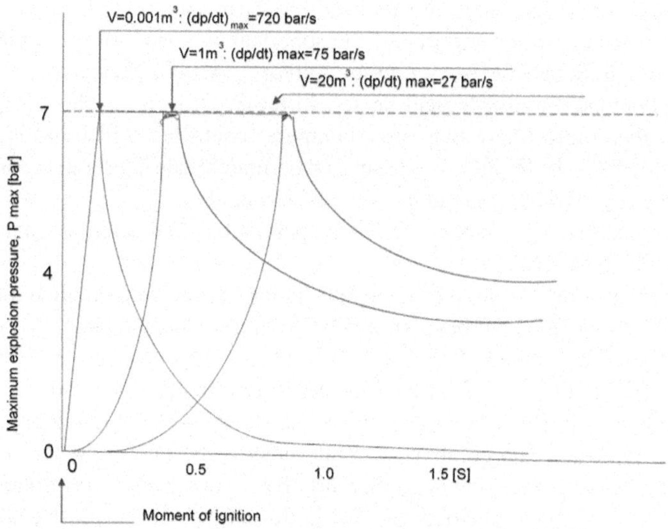

Fig. 3.2 Maximum explosion pressure developed in a closed vessel

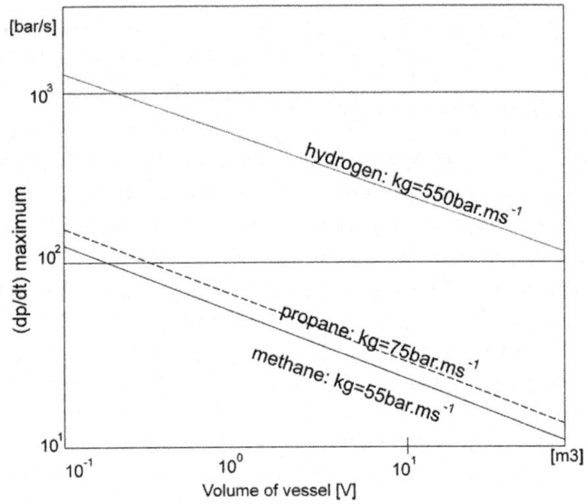

Fig. 3.3 Variation of maximum rate of change of pressure with volume of closed vessel

where k is the specific constant ratio for the

- Same ignition source.
- Same shape of the reaction vessel.
- Same turbulence of the gas–air mixture.
- Same optimum concentration of the gas–air mixture.

As a result, it is not sufficient to quote the maximum rate of pressure rise without also providing the volume.

When an enclosure has the same volume but a different shape, such as a rectangular enclosure and a spherical enclosure, then the surface area of the rectangular enclosure will be larger than the surface area of the spherical enclosure of the same volume. This means that the rate of rise of pressure $(dp/dt)_{exp}$ will be greater in a rectangular enclosure compared to a spherical vessel. In the event of an explosion within the enclosure, it is more likely that the rectangular enclosure will be broken than a spherical vessel. It is for this reason that when manufacturing flameproof junction boxes, the corners of the enclosure should be rounded.

The flamepath and gaps are two essential parameters that determine a flameproof product's flameproofness.

As shown in Fig. 3.4, as the flamepath increases, the Maximum Experimental Safe Gap also increases, but beyond 0.75″ or 19 mm flamepath, there is no effect on the Maximum Experimental Safe Gap.

Reduced flange width of flameproof enclosures means reduced resistance to outflowing gases. It is believed that as resistance decreases, flame velocity increases during the explosion and combustible materials escape into the outer atmosphere with greater velocity, resulting in greater turbulence (in the outer atmosphere). There will be a shorter contact time between the flame and the outer explosive atmosphere

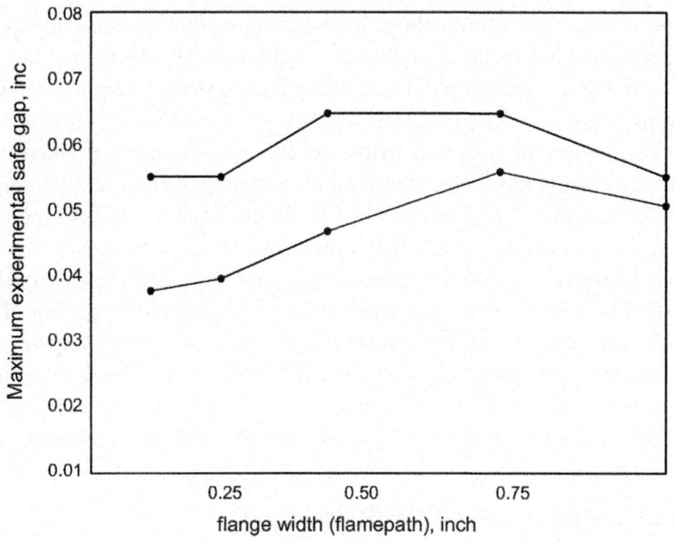

Fig. 3.4 Variation of the Maximum Experimental Safe Gap (MESG) with respect to the flamepath

due to the increased velocity. Therefore, the likelihood of the flame igniting the outer explosive atmosphere will be reduced.

Furthermore, it was found that ignition at the bottom of the enclosure causes less vulnerability, while ignition at the top poses the greatest risk. The most dangerous condition is when ignition occurs close to the inner edge of the flange gap. Therefore, increasing the width of the flange beyond 25 mm would not provide any benefit.

3.6 An Analysis of the Effect of Internal Components and the Ignition Source on the Development of Explosion Pressure in a Flameproof Apparatus

A flameproof electrical device requires adequate space to accommodate its components in order to maintain its mechanical integrity in the event of an explosion. It requires sufficient strength in terms of the thickness of the body and cover, the use of high-tensile bolts, or threaded attachment between two points. The majority of flameproof enclosures are designed and validated using Computer-Aided Design (CAD) software. The maximum pressure that develops in the flameproof enclosure is referred to as the reference pressure. An experiment was conducted to determine the effect of the position of the ignition source and the obstacle (internal component) on the rate of pressure rise and the maximum explosion pressure that developed within the enclosure [22].

Flameproof enclosures should be able to withstand the explosion pressure created by arcs, sparks, and hot surfaces of electrical equipment. Furthermore, the enclosure should prevent the transmission of internal flames to external flammable media. This maximum pressure developed inside the flameproof enclosure is a measure of the enclosure's strength and ability to withstand the pressure created when flammable media is ignited inside the enclosure. For the designer and end user, it is important to understand the effects of the placement of the ignition source and component inside the flameproof enclosure on the development of the maximum explosion pressure (P_{max}) and the maximum rate of pressure rise $(dp/dt)_{exp}$. As previously discussed, the maximum pressure developed inside a closed vessel is independent of its shape and size. The rate at which the pressure rises decreases with increasing vessel size. A study conducted by the author and his team at CSIR-CIMFR, Dhanbad, demonstrates how the position of the ignition source and component inside a flameproof enclosure can affect the rate at which pressure changes and the maximum pressure developed within the enclosure.

Experimental setup at CSIR-CIMFR, Dhanbad

The author and his team conducted an experiment to determine the effect of internal components and ignition sources on the development of explosion pressure in flameproof enclosures. During the experiment (10% v/v), a methane–air mixture was prepared in a gas chamber and transferred into the test enclosure, where it was ignited by a fuse wire/spark plug. The maximum explosion pressure developed inside the enclosure was measured using a piezoelectric pressure transducer and charge amplifier connected to a personal computer. During this test, fresh air–gas mixtures were tested at least three times and pressure–time histories were recorded. Figure 3.5 illustrates the data acquisition system for recording explosion pressures. Testing was conducted by placing the ignition source close to the wall/body for an empty enclosure as well as one equipped with obstacles.

For empty as well as enclosures fitted with components, the tests were conducted with the ignition source at the center and close to the wall/body. During experiment, two cylindrical enclosures were used. Diameter and height of cylindrical enclosure 1 were $d_1 = 10.5$ cm, $h_1 = 6.5$ cm and cylindrical **enclosure 2** was $d_2 = 15.5$ cm, $h_2 = 10.0$ cm, respectively. Table 3.2 presents the details.

The explosion pressure–time curves were drawn for cylindrical enclosures 1 and 2 for the position of the ignition source at the center (Figs. 3.6 and 3.7) and for the position of the ignition source near the wall/body (Figs. 3.8 and 3.9). Experiments were conducted with and without components. In each enclosure, the maximum pressure (P_{max}) developed was measured at least three times.

Below are the details of the figures:

1. Figure 3.6 illustrates pressure variation over time in an empty cylindrical enclosure with an ignition source in the center.
2. Figure 3.7 illustrates pressure variation over time in a cylindrical enclosure containing obstacles with an ignition source located in the center.

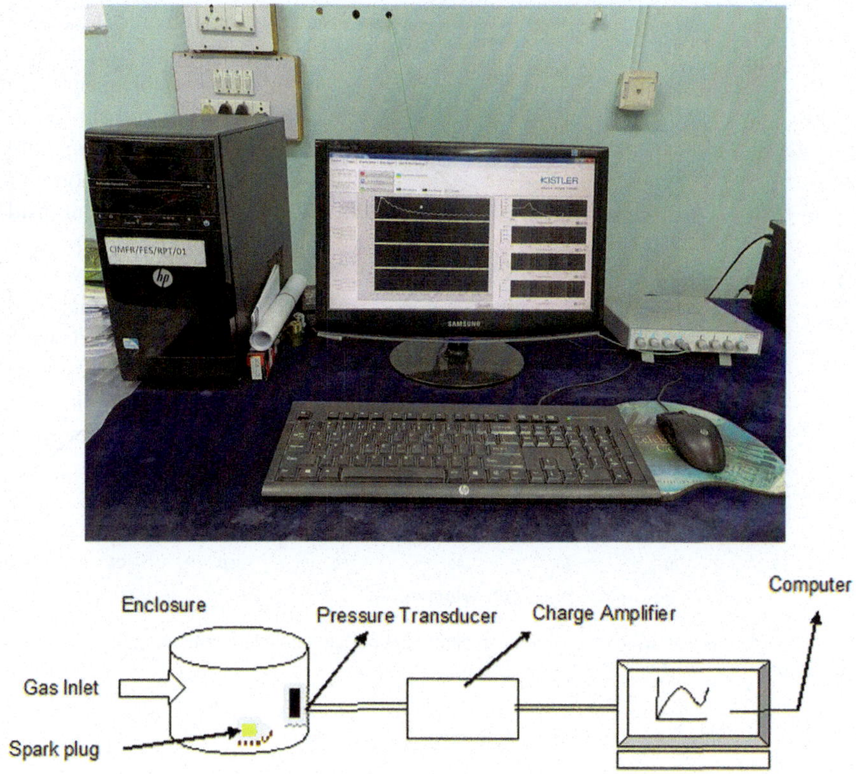

Fig. 3.5 Arrangement for recording explosion pressures

Table 3.2 Details of flameproof cylindrical enclosures with components (obstacles)

Cylindrical enclosure	Volume in liter	Component	Dimension of internal component/obstacles (cm)		
			Height	Width	Diameter/length
Enclosure 1	0.563	PBC	2.8	xx	9.5/xx
Enclosure 2	1.888	Solid metal block	6.5	5.0	xx/13.0

3. Figure 3.8 illustrates pressure variation over time in an empty cylindrical enclosure with an ignition source near the wall/body.
4. Figure 3.9 illustrates pressure variation over time in cylindrical enclosures with components with ignition sources near the walls or bodies.

Table 3.3 presents the maximum rate of pressure rise $(dp/dt)_{exp}$.

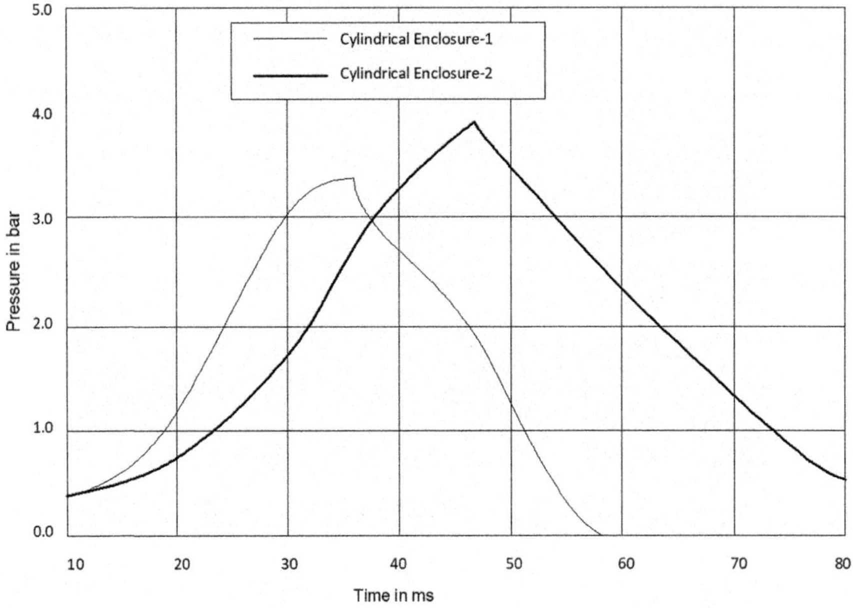

Fig. 3.6 Pressure variation over time in an empty cylindrical enclosure with an ignition source in the center

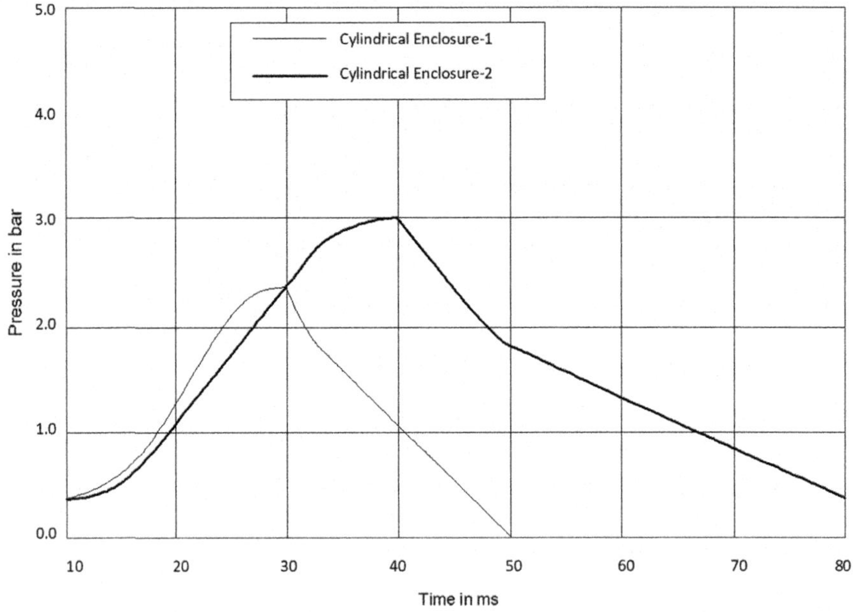

Fig. 3.7 Pressure variation over time in a cylindrical enclosure containing obstacles with an ignition source located in the center

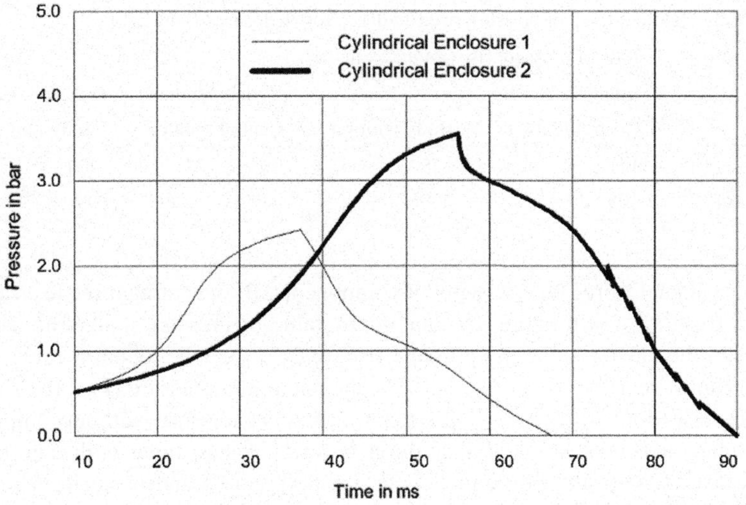

Fig. 3.8 Pressure variation over time in an empty cylindrical enclosure with an ignition source near the wall/body

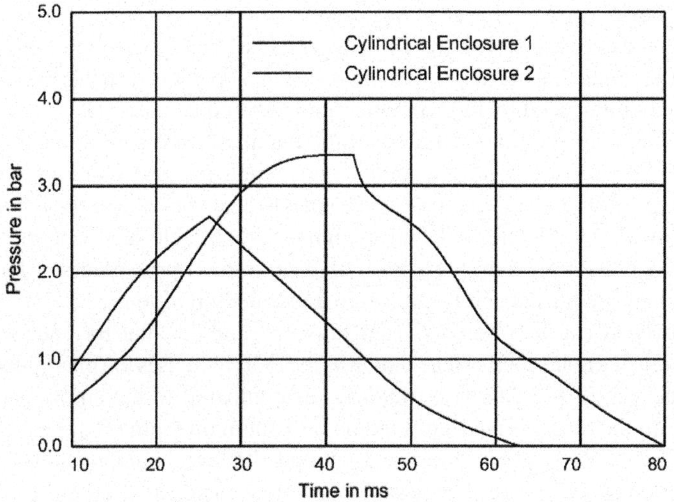

Fig. 3.9 Pressure variation over time in cylindrical enclosures with components with ignition sources near the walls or bodies

3.7 Results and Discussion

In Figs. 3.6 and 3.7, the variation in pressure is illustrated for cylindrical enclosures 1 and 2 at the ignition source at the center. Figures 3.8 and 3.9 show the position of the ignition source near the wall or body. Explosion pressures were measured in

Table 3.3 Maximum rate of pressure rise $(dp/dt)_{exp}$ for enclosures 1 and 2

Enclosure	Rate of pressure rise $(dp/dt)_{exp}$ in bar/s			
	Empty enclosure		With components (obstacle)	
	Central ignition	Body ignition	Central ignition	Body ignition
Enclosure 1	141	125	95	131
Enclosure 2	105	100	84	99

cylindrical enclosures with or without components. By repeating three tests at least with a fresh air–gas mixture, the maximum value of pressure is determined. The values of the rate of maximum pressure rise $(dp/dt)_{exp}$ can be found in Table 3.3.

As can be seen from Figs. 3.6 and 3.7, the maximum pressure (P_{max}) and rate of pressure rise $(dp/dt)_{exp}$ of central ignition in an empty enclosure (without component) are higher than those of central ignition in an enclosure fitted with components (obstacles). Due to the proximity of the ignition source and obstacle in Fig. 3.7 (central ignition with an obstacle), the pressure has developed during the explosion in this case.

It prevents rapid development and attenuates pressure waves at an early stage. Thus, the maximum pressure and rate of pressure rise recorded by the piezoelectric pressure transducer were lower in this case.

During the experiment conducted in enclosures 1 and 2 with components (obstacles), it was found that the maximum pressure developed inside these enclosures was significantly influenced by the type of component and its material.

Figures 3.8 and 3.9 demonstrate that the maximum pressure in enclosure 2 and the rate of pressure rise $(dp/dt)_{exp}$ in the empty enclosure are higher than the enclosure having a component (metal obstacle) inside enclosure 2. On the other hand, in Figs. 3.8 and 3.9, when the enclosure 1 contains an obstacle such as a printed circuit board (PCB), the maximum pressure (P_{max}) and rate of pressure rise $(dp/dt)_{exp}$ are higher than those for the empty enclosure under body ignition.

Moreover, it can be observed from Figs. 3.8 and 3.9 that for enclosure 1, the introduction of a non-metal obstacle such as the PCB increases the maximum pressure (P_{max}) and rate of pressure rise $(dp/dt)_{exp}$ as compared to the empty enclosure 1. However, in the case of body ignition, the maximum pressure (P_{max}) and the rate of pressure rise $(dp/dt)_{exp}$ for enclosure 2 with a solid metal component (Obstacle) in body ignition condition are lower than for an empty enclosure 2. (Fig. 3.9) Maximum pressures (P_{max}) and rates of pressure rises $(dp/dt)_{exp}$ are higher in enclosure 1 than in enclosure 2, for the reason that heat sinks in non-metal obstacles such as PCBs are lower in enclosure 1 than in enclosure 2. Therefore, rapid combustion occurs between the PCB as an obstacle and the wall of enclosure 1, resulting in a high explosion pressure. Based on the above results, it is evident that components and obstacles play an important role in the development of maximum pressure (P_{max}) and rate of pressure rise $(dp/dt)_{exp}$. If metal obstacles/components are involved, heat sinks will be greater, losses will be higher, and combustion will be slower with gas–air mixtures, resulting in a lower maximum explosion pressure (P_{max}) and a slower rate of pressure rise $(dp/dt)_{exp}$.

3.8 Factors Influencing Cooling of Flame and Combustible Products

It is undeniable that the cooling provided by the flange is one of the contributing factors, but it is not the only one [10]. Many other factors, as discussed below, also play a role in contributing to the cooling process.

3.8.1 Inhibition of Chain Carrier

According to experiments, quenching of the flame is also caused by the inhibition of chain carriers at the cold solid surface of the enclosure wall. In the area adjacent to the wall, a layer of the gas mixture is formed that is not traversed by the flame. As a result of heat loss to the wall, this dead layer close to the wall plays a crucial role in quenching the flame during its passage from within the enclosure to the outside.

3.8.2 Effect of Adiabatic Expansion

In adiabatic cooling, the temperature of the system is lowered without any heat or mass being transferred between it and its surroundings. According to the experiment, the flame temperature during a methane–air explosion is approximately 1800 °C. However, the mixture ignites at temperatures below 1000 °C. So, does adiabatic expansion result in a decrease in temperature from 1800 to 1000 °C. Studies indicate that adiabatic cooling does not have such significant impact on the quenching of flames, but it plays a role in quenching.

3.8.3 Influence of Lag on Ignition and Turbulence

The first event is the presence of an ignition source. The second event is the appearance of flames in a flammable atmosphere. The lag of ignition refers to the time interval between the occurrence of the first event and the occurrence of the second event. When the ignition source is a spark, the lag of ignition is short. When the ignition source is a heated surface, the lag of ignition is very long. There is a lag in the ignition even when the first event is triggered by a flame.

The speed at which flames and hot gases are projected from the enclosure's flange to the outside atmosphere is directly proportional to the explosion pressure generated within the enclosure. In the case of a low explosion pressure and a low speed, the time of contact between the burning gas and the unignited gas is long. As a result, the chances of ignition are much greater. Furthermore, if the explosion pressure and speed are high, the time that the burning gas is in contact with the unignited gas will be lesser. This reduces the chances of ignition.

3.8.4 *Effect of Flamepath and Gap*

Whatever the point of ignition, a reduction in flamepath generally results in a significant reduction in the Maximum Experimental Safe Gap (MESG). Table 3.3 indicates the variation of flamepath with Maximum Experimental Safe Gap.

A reduction in the flamepath of a given enclosure would decrease the resistance to outflowing gases. Some may anticipate that this will increase the velocity of flame projection and combustible products, thereby increasing turbulence in the outer atmosphere and reducing the possibility of fire. But that would happen only when the pressure is maintained, so in actuality, a smaller flamepath leads to a lower explosion pressure within the enclosure, resulting in a reduced velocity of projection of flame and combustible products and a greater likelihood of ignition, resulting in a lesser MESG (Table 3.4).

In addition, it is imperative to keep in mind that increasing the flamepath beyond a certain point does not yield the desired results. Various experiments have demonstrated that increasing flamepath beyond 0.75 inches has no effect on MESG, as shown in Fig. 3.4.

3.8.5 *The Effect of the Point of Ignition*

The point of ignition is also an influencing factor in determining the transmission of flame and combustible products. There has been evidence from a variety of experiments that

1. Ignition at the bottom of the enclosure is less vulnerable.
2. Ignition at the top of the enclosure is most vulnerable.
3. Ignition located near the inner edge of the flange gap is the most dangerous.

Table 3.4 Influence of length of flamepath (L) on the maximum gap (volume above 100 cm^3 and up to 2000 cm^3)

Flamepath (mm)	Maximum gap or diametrical clearance (mm)		
	Gr. I	Gr. IIA	Gr. IIB
Flange and spigot joints			
$L \leq 25$	0.5	0.4	0.2
$*L \leq 12.5$	0.4	0.3	0.2
Operating rods and spindles			
$L \leq 25$	0.5	0.4	0.2
$L \leq 12.5$	0.4	0.3	0.2
Shafts with sleeve bearing			
$L \leq 40$	0.6	0.5	0.3
$L \leq 25$	0.5	0.4	0.25
$L \leq 12.5$	0.4	0.3	0.2
Shafts with ball and roller bearing			
$L \leq 40$	0.75	0.75	0.45
$L \leq 25$	0.6	0.6	0.4
$L \leq 12.5$	0.6	0.45	0.3

*L—maximum length of flamepath

References

1. Chakraborty N, Scheid A (2014) History of development of FLP /IS electrical and emerging trends in development. In: 2nd international seminar & exhibition on "Recent trend in design, development, testing and certification of ex-equipment for explosive atmosphere, 7–9 Nov 2014, at Kolkata, India
2. BS 229-1926 for certification flameproof equipment
3. Coal Mine Act 1911
4. BS 1259: 1945 Intrinsically safe electrical apparatus and circuits for use in explosive atmospheres
5. BS 1259: 1958 Intrinsic ally safe electrical apparatus and circuits for use in explosive atmospheres
6. 1st NEC for IS testing in 1965 (NEC 504:1965)
7. IS: 5780 – 1971 Intrinsically Safe apparatus and circuits for use in explosive atmosphere
8. IS 2148:1968 Flameproof enclosure for electrical apparatus for use in explosive atmosphere
9. IEC 60079-20:1 (2013) Electrical apparatus for explosive gas atmospheres, part 20: material characteristics for gas and vapour classification
10. Sarkar SK (1988) Flameproof equipment, design, construction, certification use and maintenance, 1st eds. (cientific Mining Publisher, Dhanbad, pp 34–36 (May 1988)
11. Nagy Y, Seiler E, Conn J, Verakis H (1971) Explosion development in closed vessels. US Department of the Interior (USBM)-Report of Investigations 7505
12. Harris GFP, Briscoe PG (1967) The venting of pentane vapour air explosions on a large vessel. Combust Flame 11:329338
13. Fan Y, Crowl DA (2000) Predicting the maximum gas deflagration pressure over the entire flammable range. J Loss Prev Process Ind 13:361368
14. Amyotte PR, Patil S, Pegg MJ (2002) Confined and vented ethylene air deflagrations at initially elevated pressures and turbulence levels. Process Saf Environ Prot 80(Part B):7177

15. Razus D, Movileanu C, Brinzea V, Oancea D (2006) Explosion pressure of hydrocarbon air mixtures in closed vessels. J Hazard Mater 135(13):5865
16. Gieras M, Kelmens R (2009) Experimental studies of explosions of methane air mixtures in a constant volume chamber. Combust Sci Technol 181:641653
17. Phylaktou H, Andrews G (1991) Gas explosions in long closed vessels. Combust Sci Technol 77:2739
18. Cashdollar K, Zlochower I, Green G, Thomas R, Hertzberg M (2000) Flammability of methane, propane, and hydrogen gases. J Loss Prev Process Ind 13:327340
19. Dahoe A, de Goey L (2003) On the determination of the laminar burning velocity from closed vessel gas explosions. J Loss Prev Process Ind 16:457478
20. Andrews G, Herath P, Phylaktou H (1990) The influence of flow blockage on the rate of pressure rise in large L/D cylindrical closed vessel explosions. J Loss Prev Process Ind 3:291302
21. Gieras M, Klemens R, Rarata G, Wolanski P (2006) Determination of explosion parameters of methane air mixtures in the chamber of 40 dm^3 at normal and elevated temperature. J Loss Prev Process Ind 19:263270
22. Vishwakarma RK, Ranjan V, Kumar J, Singh AK (2014) Influence of position of ignition source and internal component on explosion pressure development in flameproof enclosure. In: 2nd international conference on design, development, testing and certification of ex-equipment, 7–9 November, 2014 at Kolkata, India, pp 111–117

Chapter 4
Concept of Protection Against Explosion in Flammable Dust Atmosphere

4.1 Introduction

A hazardous area is not only created by the presence of flammable gas, vapor, or liquid but also by the presence of combustible dust. For example, coal dust in underground coal mines. During the extraction of coal from underground coal mines, loading and unloading coal, or transporting it from one place to another, combustible coal dust settles on the surface of electrical equipment. This forms a thick layer of coal dust that increases the surface temperature of the electrical product and violates the certified temperature class of the electrical product. Due to the settlement of coal dust, the surface temperature of electrical equipment increases and exceeds the auto-ignition temperature of a coal dust cloud, resulting in an explosion that may be the cause of unfortunate incidents in coal mines below ground. In refineries like IOCL Panipat, purified terephthalic acid (PTA), dust is generated during the refining of crude oil and can be ignited if it comes into contact with a source of ignition. The dust may be produced either as an end product or as a by-product. The end products include coal dust, plastic dust, metal dust, sugar, flour, starch, and pharmaceutical dust, while by-product dust includes wood sawdust, textile fluff, and dust generated during the cleaning or polishing of solid surfaces. The handling of such type of combustible dust in the industry is a risky activity. This is because, when it comes into contact with an ignition source, a fire or explosion may result. Dust fires generally develop in layers, storage bins, heaps, and other deposits [1]. If suspended dust is ignited, it causes an explosion. When burning dust particles fall out of suspension onto nearby surfaces, the fire spreads. In order to protect against explosions, it is necessary to design safe electrical products that are reliable and economical.

© The Author(s), under exclusive license to Springer Nature Singapore Pte Ltd. 2023
A. Kumar Singh, *Explosion-Proof Equipment in Hazardous Area*,
https://doi.org/10.1007/978-981-99-2516-2_4

In this chapter, the author tries to discussed about mechanism of dust explosion, classification of extent of hazard due presence of combustible dust, industries that produce ignitable dust, source of ignition, suitable apparatus for use in dusty atmosphere, mechanism of calculation of maximum surface temperature of electrical equipped in dusty atmosphere, variation of minimum ignition energy of coal dust at varying temperatures and minimum ignition temperature at various thickness layer of coal dust.

4.2 Dust Explosion

Dust explosions and fires occur in many industries, but the literature on this subject is not unified. There are many solid materials that are combustible in nature and can explode if an ignition source is present. Solid dust material burns in accordance with its dimensions. Combustible solids burn at a higher rate with increasing subdivision because the total contact surface area between the solid and air increases [2]. Consequently, heat is liberated at a high rate as a result of the following reaction: fuel + oxygen = oxides + heat.

There are several groups of materials that can cause dust explosions: natural organic materials (grain, wood, linen, sugar, etc.), synthetic organic materials (plastics, organic pigments, pesticides, pharmaceuticals, etc.), coal, peat, and metals (aluminum, magnesium, titanium, zinc, iron, etc.).

The dust cloud can ignite and burn more violently when the dust particles are small, up to a certain particle size that depends on the dust material. This type of explosive combustion can occur inside process equipment or workrooms, and the explosion space may burst, resulting in the loss of life and property (Fig. 4.1).

4.3 History of the Dust Explosion

Dust explosions have probably existed since man developed windmills to grind grain. An account of a violent explosion of flour dust in a Turin mill in 1785 is the earliest known report of a dust explosion [3]. The dispersed dust was believed to have produced an inflammable gas, which was then ignited. Table 4.1 shows the some dust explosion incidents across the Globe.

Explosions in coal mines were common during this period, but no consideration was given to the possible effects of airborne coal dust. In the last half of the nineteenth century, it was realized that coal dust could ignite in the absence of gas, and that flour dust alone could cause mill explosions.

New materials and industrial development led to increased dust explosions. In 1919, 43 workers at an Iowa, USA starch factory were killed in a dust explosion [4]. Explosions of dust prompted research. 18 deaths occurred in a Scottish mill in 1872 and six in a French dye factory in 1878, prompting independent investigations in the

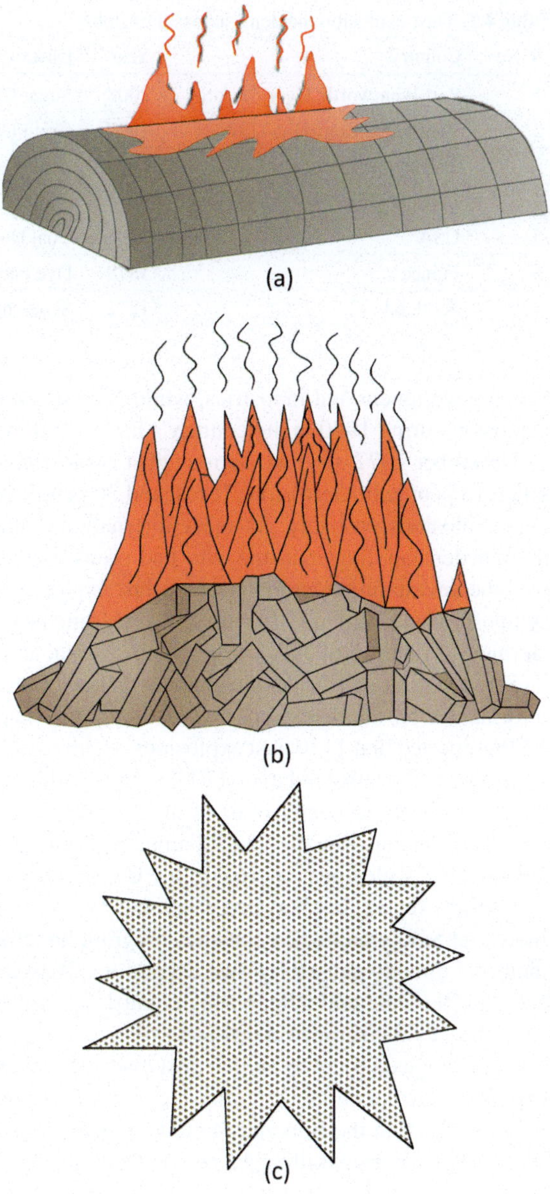

Fig. 4.1 a–c Combustion rate increases with increasing the sub-division of combustible solid

USA, Great Britain, and France. Dust explosions contributed to 1148 coal-mining deaths in 1907. In 1908, the US Bureau of Mines was established to study the causes of coal dust explosions. In more recent times, a sugar dust explosion in a Swedish silo in 1966 resulted in the Dust Research Institute in Bonn undertaking an investigation of silos; an explosion in a Dutch fodder plant led to research being sponsored by the

Table 4.1 Dust explosion incidents across the Globe

S. No	Country	Year	Dust explosion	Casualty/death
1	Port Wentworth, Georgia, USA	2008	Sugar Company Plant	12
2	Westwego, Louisiana USA	1977	Grain Dust Explosion	36
3	Galveston, Texas, USA	1977	Grain Dust	18
4	Lowa USA	1919	Starch Factory	43
5	USA	1907	Coal Dust	1148
6	France	1878	Dye Factory	6
7	Scotland	1872	Scottish Mill	18

Dutch government and industries, while Norwegian research was stimulated by an explosion during 1970 in the country's largest and most modern import grain silo.

December 1977 saw two serious dust explosions. The explosions of grain dust killed 18 people in Galveston, Texas, and 36 people in Westwego, Louisiana. These dust explosions probably got more international attention. Many professionals in the powder handling industry who had previously ignored the problem now realized the devastating consequences of dust explosions. In response, an international symposium was held in 1978 with speakers from the USA, UK, Canada, Switzerland, Germany, France, Holland, USSR, South Africa, and Australia.

Statistical data on dust explosions are difficult to obtain, and they can only provide information on incidents reported to the appropriate authorities. In the USA, the NFPA reported that 1120 dust explosions occurred between 1900 and 1956, resulting in at least 640 deaths and over 1700 injuries. Additionally, it has been reported that there have been 220 explosions in the American grain industry between 1958 and 1977, resulting in 48 deaths and 500 injuries. Between 1900 and 1956, approximately, 50% of dust explosions reported in the USA occurred in grain-handling processes.

Griffiths report that between 1969 and 1973, the Japanese Research Institute of Industrial Safety published statistics on grain-handling plants. This period saw 187 incidents directly caused by dust explosions. Between 1960 and 1972, the Federal German Republic and neighboring countries experienced at least 4000 dust explosions, based on insurance claims, an average of one dust explosion per working day. Most people would not think that dust made of ordinary material such as food could explode but under certain circumstances, it can. A violent explosion occurred on 7th February 2008 at the Imperial Sugar Company Plant on the Savannah River at Port Wentworth, Georgia, killing 12 people and injuring 15.

4.4 How is the Behavior of Combustible Dust Different from Gases and Vapors

According to NEC, dust cloud processing requires a high degree of control to prevent disasters. Unlike gas/vapors, dust settles and forms layers on the surface of the enclosures. The solution to this problem lies in selecting the electrical apparatus based on the zone.

Zone 20 is defined area in which there is a normal presence of dust clouds, such as the interior of a vessel. It is similar in definition to zone 0 for gases and vapors.

Zone 21 is defined as an area in which dust clouds are frequently produced by controlled sources, such as loading apertures. Its definition is similar to that of zone 1 for gases and vapors.

Zone 22 can be defined as an area where there is hardly any presence of combustible dust clouds. However, it can be found if dust layers are disturbed, just as zone 2 applies to gases and vapors. As a result, dust hazards will only last for a short period of time in such situations.

According to IEC 60529 [5], only products in the Intrinsically Safe category 'ia' with a degree of protection IP6X are permitted for zone 20.

4.5 Industries that Produce Ignitable Dust

It is important to note that there are numerous industries that produce explosive dust. The followings are some of the most important industries:

- Woodworking
- Plastics
- Agriculture
- Coal, mining, and utilization
- Foodstuffs, human, and animal
- Chemicals, including dyestuffs
- Pharmaceuticals
- Metals.

4.6 Source of Ignition

The industrial plant should be designed in such a manner as to minimize the possibility of ignition sources. However, in practice, it is not possible to completely eliminate the generation of dust. Therefore, it is imperative that the sources of ignition be minimized to the greatest extent possible. A large number of ignition sources exist, each with a different energy level and temperature. These are as follows [1]:

- Spontaneous heating
- Electric sparks
- Frictional and impact
- Flames
- Hot surfaces
- Welding and cutting
- Glowing combustion and smoldering.

4.7 Dust Fires

In a dusty atmosphere, fire is caused by the combustion of combustible dust and oxygen. Dust may be deposited in layers, heaps, or other forms. Dust fires and explosions are different. Fires spread slower than dust explosions.

4.8 Apparatus for Combustible Dust

Depending on the degree of enclosure, there are two kinds of apparatus.

1. **Dust-tight apparatus**—Describes an enclosure that is totally dust-tight and capable of excluding all observable dust. A degree of protection IP_{6X} is required for the apparatus (i.e., protection from dust and liquid) where 'X' lies between 0 and 8 according to IEC 60529:2001 [5]. Zone Z or zone 21 requires equipment with an IP_{65} enclosure. The highly flammable category includes metal dust, such as magnesium powder and coal dust.
2. **Dust-protected apparatus**—A dust-protected apparatus cannot prevent the ingress of dust, but this does not affect the normal operation of the equipment. The apparatus having a degree of protection against dust and liquid, namely IP_{5X} as per IEC 60034: Part 5 [6], may be installed in zone 22/zone Y. This enclosure will be suitable for dust that is non-conductive, such as wheat flour. In terms of dust mass, the maximum amount that can enter is equal to 10 g per cubic meter divided by the enclosure's free volume.

4.9 Functional Requirement/Performance Requirement

Sparks may or may not occur within the enclosure, as the enclosure provides protection. Therefore, there is no need to limit the energy of a spark or arc but only to limit the maximum external surface temperature of the enclosure. In order to determine the enclosure temperature, the enclosure should be mounted in the most unfavorable normal mounting position (the one that produces the highest temperature on any part of the enclosure's surface); second, the supply voltage must be adjusted to any value within its operating tolerance to give the maximum temperature in the above position (if the supply voltage tolerance is not quoted, a 10% tolerance is typically used).

The maximum surface temperature of electrical equipment used in dusty environments is 10 °C less than the allowed value for dust clouds [7]. This can be calculated as follows:

1. The maximum surface temperature of electrical equipment that can be used in a dusty cloud of sulfur having a cloud ignition temperature of 280 °C is [(280 × 2/3) – 10] = 176 °C at 40 °C ambient temperature (A surface temperature elevation of 136 °C at an ambient temperature of 40 °C).
2. Let us assume that the ignition temperature of a 5 mm thick dust layer is 225 °C. The maximum surface temperature of electrical equipment/enclosure that can be permitted is (225–75–10) °C = 140 °C.

In accordance with IEC 60079-0:2017 [8], a margin of 10 °C is allowed for equipment normally found in dusty environments. This is the reason, the dust cloud ignition temperature is multiplied by two-thirds and subtracted from its layer ignition temperature to identify the suitability of electrical equipment used in a dusty atmosphere. The electrical equipment which certifies the above-mentioned temperature condition may operate in a gassy and vapor atmosphere. All types of protected electrical equipment may be used in such risky atmospheres except the non-sparking (Ex 'n') type of protection. Non-sparking products are subject to a temperature relaxation of 5 °C for (T3–T6) temperature classes and 10 °C for T1 and T2. Table 4.2 shows the minimum dust ignition temperature corresponding to the classification of temperature class [7].

Table 4.2 Minimum combustible dust ignition temperature with respect to temperature class

T-class	Combustible dust ignition temp			
	Method of protection Ex 'd', Ex 'e', Ex 'i', Ex 'm', Ex 'o', Ex 'p' and Ex 'q'		Method of protection Ex 'n'	
	Dust cloud (°C)	Dust layer (°C)	Dust cloud (°C)	Dust layer (°C)
T1	675	525	685	535
T2	450	375	465	385
T3	308	280	315	285
T4	195	205	203	210
T5	158	180	165	185
T6	135	165	143	170

4.10 Requirement for Dust Ignition Protection

Combustible dust can be easily ignited if it comes under contact with hot surface and below that temperature will be the permitted surface temperature of the electrical equipment intended for use in a dusty atmosphere. Therefore, the surface temperature of any electrical equipment used in a dusty atmosphere needs to be established. Many times, it has been observed that the lowest ignition temperature of dust is calculated when the surface of the electrical equipment is covered with a 5 mm thick layer of combustible dust.

However, the ignition temperature is dependent on the nature and size of the dust particles. Finally, this cannot be stated with certainty until these tests have been conducted by a competent laboratory. A protected product designed for use in a gassy atmosphere may also be used in a dusty atmosphere, such as zone 20, zone 21, and zone 22. The following parameters should be considered in such cases [10].

4.10.1 Type of Method

Various method of protection is prevalent and is suitable for use in hazardous areas depending on extent of hazard, such as flameproof, intrinsically safe, increased safety, purge protected, encapsulation, non-sparking, oil filled, and sand filled.

Flameproof Ex 'd' equipment for zone 1 and zone 2, intrinsically safe Ex 'i' for zone '0', zone 1 and zone 2, oil filled Ex 'o', and increased safety Ex 'e' for zone 2 or any other type of protection.

4.10.2 Ingress Protection (IP)

Ingress protection means degree of protection with respect to ingress of dust and liquid. Its first digit indicates the ingress of dust, and second digit indicates the ingress of liquid (water). Although electrical equipment may be used in a gassy flammable atmosphere if it is tested for dust and liquid protection, it is not a final acceptance criterion for use in a dusty combustible environment since zone z/zone 21 areas require an IP 6X rating, whereas zone y/zone 22 requires an IP 5X rating.

4.10.3 Maximum Surface Temperature

The auto-ignition temperature of the flammable gases should always be lower than the surface temperature of the electrical equipment installed in a dusty environment. A suitable safety margin has been included in the calculation.

It is recommended that the maximum surface temperature of the enclosure or electrical equipment be less than the following:

- The maximum surface temperature of the enclosure should not exceed two-thirds of the minimum ignition temperature of the combustible dust or dust layer.
- If the combustible dust layer thickness is 5 mm, then the enclosure's surface temperature must be less than 75 K from the ignition temperature of the combustible dust layer.

4.10.4 Minimum Ignition Temperature of Dust Cloud

Generally, unwanted dust clouds are generated when powder/dust material falls freely under gravity. Therefore, whenever possible, it is advisable to design transfer points so that the material flows smoothly in bulk rather than being dispersed as a cloud. The minimum ignition temperature of a dust cloud refers to the temperature at which it will automatically ignite when exposed to the heated air in a furnace under local atmospheric pressure.

Based on the mixing of different dusts, the following observations have been made regarding the minimum ignition temperature of dust clouds [11]

1. As the mass fraction of inert dust increases in a mixture containing combustible (flour) and inert dust, the minimum ignition temperature of coal dust decreases.
2. The minimum ignition temperature of coal dust is likely to increase as inert dust undergoes endothermic decomposition.
3. There is a limited effect of limestone on the minimum ignition temperature of coal dust: When 10% flour is added to coal dust, the temperature only rises to 540 °C.

4.10.5 Hazard Possibilities

The followings are the main factors which creates hazard.

a. Risk of toxic gases.
b. Risk due to fire.
c. Due to static charge developed on the enclosure of electrical equipment.
d. Due to thundering and electro-magnetic rays.
e. Source of ignition due to frictional non-incentive properties of material of construction of the enclosure.

4.11 Ignition Protected by Product

Dusty atmospheres such as zone 21 and zone 22 require equipment to be tested and certified by a testing laboratory and meet the following requirements:

4.11.1 Design and Construction of Dust-Protected Products

In order for electrical equipment to be used in a dusty environment, it must meet the following design criteria.

1. Equipment should be designed in such a manner as to minimize the accumulation of dust on the surface.
2. There should be a minimum accumulation of static charge.
3. Easy to clean equipment.

4.11.2 Material of Construction (MOC)

Construction materials should possess the following characteristics.

1. Enclosure materials should be mechanically strong and eco-friendly. It should be constructed of materials capable of passing impact and drop tests in accordance with the relevant International Electro-technical Commission (IEC) standards.
2. During adverse conditions, the construction material of the equipment should be such that it can withstand a maximum temperature rise.
3. In the case of equipment manufactured from LM6 material, the concentration of magnesium should not exceed 7.5% in the alloy.
4. Non-metal construction materials should be non-flammable and non-combustible.
5. It is critical to keep in mind that dusty hazardous areas require equipment to be certified by the Competent Authority. There must be no violation of the classification issued by the test laboratory with regard to the temperature class of the apparatus.

4.11.3 Joints and FlamePath (Dust Path) with Gap

In order to prevent dust ingress into electrical equipment, the dust path and gap between the electrical equipment and the gasket should be maintained in accordance with the relevant IEC Code. The following joints are permitted of the electrical equipment used in dusty atmosphere.

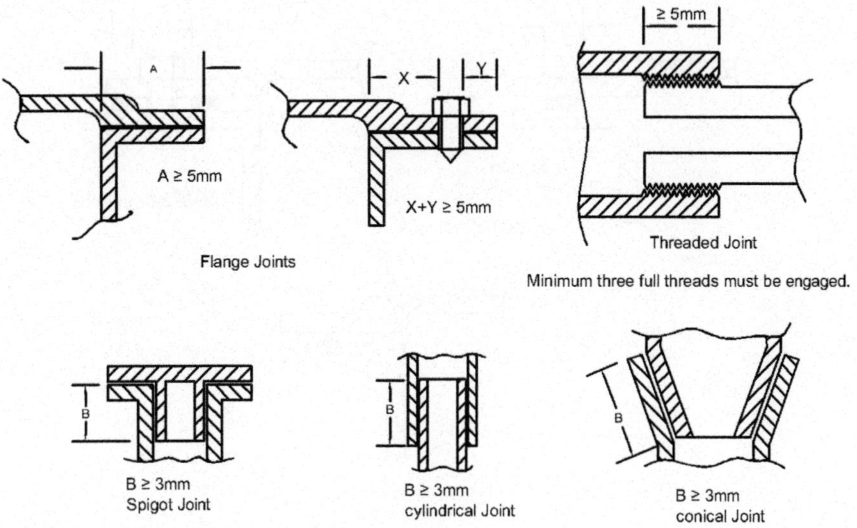

Fig. 4.2 Ungasketed joint with minimum dimension

4.11.3.1 Joints Without Gasket

Electrical equipment may be used in dusty atmospheres without a gasket if the equipment having desired IP-rating. All the flameproof equipment satisfies the condition of dusty atmosphere requirement. Therefore, joint with maintaining a minimum of 6.3 μm roughness may be applied in dusty atmosphere (Fig. 4.2).

Minimum dust ingress paths for different joints

Flanged joint ≥ 5 mm
Spigot joint ≥ 3 mm
Cylindrical joint ≥ 3 mm
Conical joints ≥ 3 mm
Threaded joints ≥ 5 mm.

To achieve the required ingress protection (IP), a non-setting sealing compound should not be used. In the case of the threaded joint, at least five threads should be fully engaged where they are parallel or taper threads. Purchaser of the equipment must be aware that no compound is used to maintained the required ingress protection (IP) of the products.

4.11.3.2 Joints with Gasket

Generally, gaskets are used in case of the flanged joint to maintain the required ingress protection (IP) as per Fig. 4.3 provided that the gasket width is at least 3 mm.

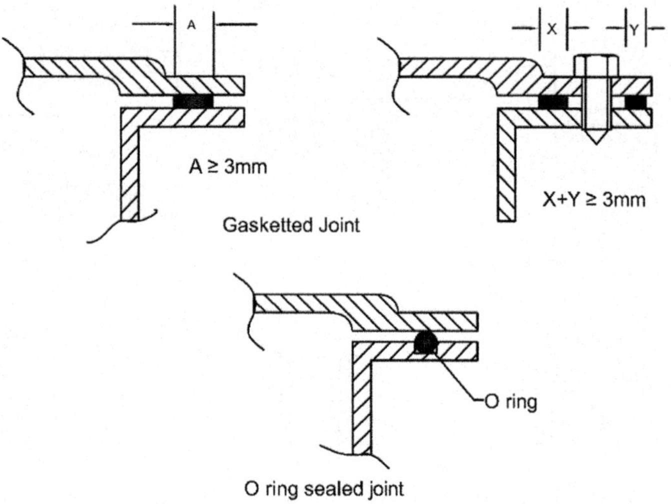

Fig. 4.3 Use of gasket and O-ring

Gaskets are generally fixed in flanged portion with adhesive to prevent its displacement but adhesive is prepared by such a material which do not stick to mating faces. This adhesive material does not damage during external use of equipment. Lip seals are used when 'o' ring is used to maintained the require degree of protection.

4.11.4 By Fastening Methods

Fasteners plays an important role to secure the part of the enclosure and protection concept. Hinged lids are applicable provided that it does not apply unnecessary pressure for sealing. When fasteners are used in electrical equipment, dust and liquid protection must be maintained. If a fastener passes through the wall of the electrical equipment, it will create a permanent pathway for the dust to enter the product. It is recommended that the wall thickness of electrical equipment be at least 3 mm or one-third of the hole diameter, whichever is greater.

4.11.5 Light Transmitting Parts

Cement or a similar method is used to fix the light transmitting part of the enclosure. The glass part should have passed the impact and thermal shock tests. It is important to maintain the Ingress Protection rating of the enclosure with glass as determined by the test laboratory. The cement used for fixing the glass should retain its properties throughout its life. A thermal shock test should be performed on glass that is used for light transmission by spraying a 1 mm jet of water at a temperature between 10 and 15 °C when the glass is at its maximum operating temperature.

4.11.6 Terminal and Conductors

In accordance with relevant IEC standards, a minimum clearance distance and creep age distance should be maintained between terminals and conductors, and both should be conducing and vibration-resistant.

4.11.7 Cable and Conduit Entries

It is recommended that cables and conductors used in hazardous areas should be entered through weatherproof double compression cable glands that have been tested and certified. Plugging unused cable entries with a certified stopping plug will maintain the threaded joint or spigot joint dust path.

4.11.8 Shaft and Spindle

If electrical equipment having shafts and spindles which rotate and move should be sealed by suitable bushing to prevent the ingress of dust, normally cylindrical joints are joint fit for such use. Joints should be such that it prevents the ingress of dust when grease or oil is absent in joint due to operational reason. For such type of joint, minimum dust path is not define in any code.

4.11.9 Maximum Surface Temperature (MST) of Enclosure

Electrical equipment used in a dusty atmosphere should always have a maximum surface temperature that is lower than the auto-ignition temperature of the combustible dust. Testing authorities must measure the surface temperature (with specified tolerance in rating) under the most adverse conditions possible. Table 4.3 shows the requirement of method of protection and corresponding extent of hazard.

Table 4.3 Requirement of dust-protected electrical products

Type of method	Extent of hazard	Requirements
Ex 'i' with 'ia' category	Zone 20	i. Minimum degree of ingress protection, IP6X ii. Temperature class of apparatus should not be greater than two-third of the ignition temperature of dust cloud iii. Temperature class of 5 mm thick dust layer should be greater than (temperature class of electrical equipment +75 K)
Ex 'i' Ex 'd' Ex 'e' Ex 'p' Ex 'o' Ex 'm'	Zone 21 and zone 22	i. If the hazardous area is zone 22, then ingress protection IP 5X and in case of zone 21, IP6X ii. Temperature class of apparatus should not be greater than two-third of the ignition temperature of the dust cloud iii. Temperature class of 5 mm thick dust layer is greater than (temperature class of electrical equipment +75 K)

4.12 Minimum Ignition Temperature of Coal

Figure 4.4 shows the minimum ignition energy of coal dust of various volatile contents decreases as temperature increases, but after reaching the minimum value, it rapidly increases [1].

Minimum ignition temperature of coal dust depends upon the thickness of dust layer. Figure 4.5 shows that minimum ignition temperature of coal dust decreases with increasing the dust thickness layer [1].

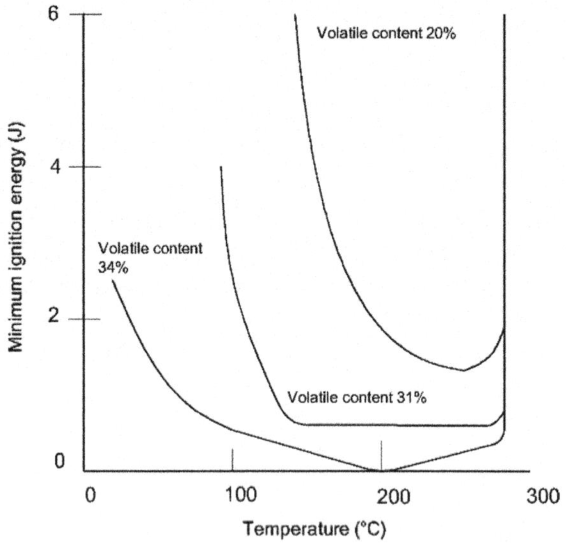

Fig. 4.4 Minimum ignition energy of coal dust at varying temperatures

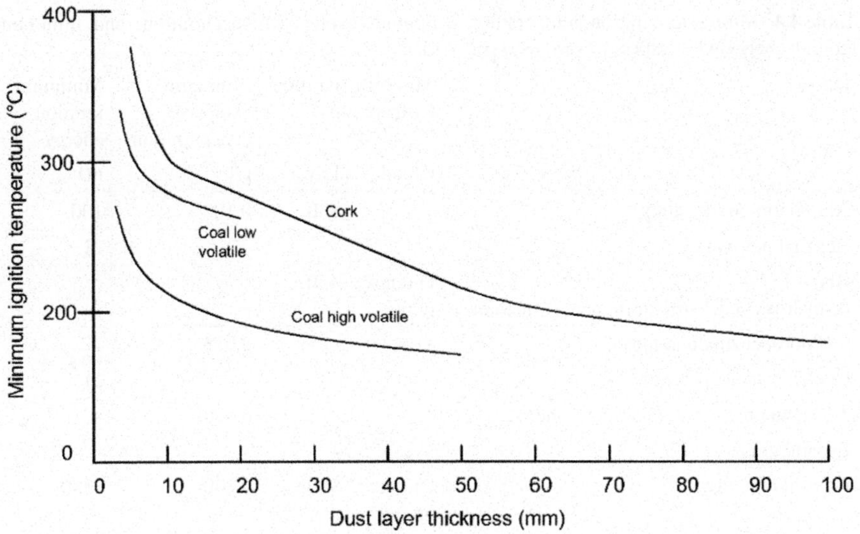

Fig. 4.5 Minimum ignition temperature of dust layer

Considering the above discussion, it is clear that dust ignition-protected products are essential for safe handling and processing in industries such as underground coal mines, oil refineries, and many other dusty industries. A non-compliant electrical product selected for a dusty environment can have a negative effect on production and may result in accidents.

Table 4.4 provides values based on materials and available data related to fire and explosion properties. This table contains values taken from publications published in the UK, USA, and Germany [1].

Table 4.4 Minimum ignition temperature, minimum explosive concentration, and minimum ignition energy with respect to various types of dust

Dust	Minimum ignition temperature °C		Minimum explosive concentration	Minimum ignition energy
	Cloud	Layer	g/l	mJ
Aluminum-copper alloy	–	830	0.100	100
Benzoyl peroxide	–	–	–	21
Bis (2-hydroxy-3,5,6,-trichlorophenyl)-methane	Did not ignite	450	–	–
o-Chlorobenzmalono nitrile	–	–	0.025	–
Cyclohexanone peroxide	–	–	–	21
3-3' Dimethoxy 4-4' diamino diphenyl	–	–	0.030	–
Esparto grass	–	–	–	–
Ferrosilicon (90% Si)	Did not ignite	980	0.240	1280
Grass	–	–	–	–
Horseradish	–	–	< 0.100	–
Lauryl peroxide	–	–	–	12
Nitrocellulose	–	–	–	30
Rayon, flock	–	–	0.03	–
Sebacic acid	–	–	–	–
Silicon	Did not ignite	760	< 0.10	80
Sodium m-nitrobenzene sulfonate	–	–	–	–
Sodium m-nitrobenzoate	–	–	–	–
Sodium pentachlorophenate	Did not ignite	360	–	–
Trinitro toluene	–	–	0.070	75
Uranium	20	100	0.060	45
Uranium hydride	20	20	0.060	5
Zirconium	20	220	0.045	5
Sulfur	190	220	0.035	15
Nitro furfural semi carbazone	240	–	–	–
Thorium hydride	260	20	0.080	3
Waz, accra	260	–	–	–
Manganese ethylene bis dithiocarbamate	270	–	0.07	35
Thorium	270	280	0.075	5
Ferric dimethyl dithiocarbamate	280	150	0.055	25
Phosphorus pentasulfide	280	270	0.050	15

(continued)

Table 4.4 (continued)

Dust	Minimum ignition temperature °C		Minimum explosive concentration	Minimum ignition energy
	Cloud	Layer	g/l	mJ
α, α'-Trithiobis (N,N-dimethyl-thioformamide)	280	230	0.060	35
Stearic acid	290	–	–	25
2,2-Methylene bis-4-ethyl-6-t-butyl phenol	310	–	–	–
Penicillin, N-ethyl piperidine salt of	310	–	–	–
Zinc stearate	315	Melts	0.020	10
Sodium carboxymethyl cellulose	320	–	1.10	440
Tobacco, dried	320	–	–	–
Colophony	325	Melts	–	–
Copal	330	Melts	–	–
Phytosterol	330	Melts	0.025	10
Cellulose acetate	340	–	0.035	20
Chocolate crumb	340	–	–	–
Dinitrocresol	340	Melts	0.030	–
Ethyl cellulose	340	330	0.025	15
Wax, carnauba	340	–	–	–
Wax, paraffin	340	–	–	–
Dextrose monohydrate	350	–	–	–
Polyethylene oxide	350	–	0.030	30
Rubber, crude, hard	350	–	0.025	50
Tartaric acid	350	–	–	–
Zirconium hydride	350	270	0.085	60
Brunswick green	360	–	–	–
Coffee	360	270	0.085	160
Garlic	360	–	0.10	240
Gum manila (copal)	360	390	0.030	30
Methyl cellulose	360	340	0.030	20
p-Nitro-benzene arsonic acid	360	280	0.195	480
Phosphorus, red	360	305	–	–
Rubber, vulcanized	360	–	–	–
Wood	360	–	–	–
Barley	370	–	–	–

(continued)

Table 4.4 (continued)

Dust	Minimum ignition temperature °C		Minimum explosive concentration	Minimum ignition energy
	Cloud	Layer	g/l	mJ
Cellulose acetate butyrate	370	–	0.025	30
Copper-zinc, gold bronze	370	190	1.00	–
Ferric ferrocyanide	370	–	–	–
Ferrotitanium	370	400	0.140	80
DL Methionine	370	360	0.025	35
Provender	370	–	–	–
L-sorbose	370	–	0.065	80
Sugar	370	400	0.045	30
Titanium	375	290	0.045	15
Benzethonium chloride	380	410	0.020	60
Carboxy methyl hydroxyl ethyl cellulose	380	–	0.200	960
Dinitrobenzoyl chloride	380	–	–	–
Ferrous ferrocyanide	380	190	0.400	–
Iron pyrites	380	280	1.00	8200
Potassium sorbate	380	180	0.120	60
Rubber	380	–	–	–
Vitamin B1 mononitrate	380	190	0.035	35
Wheat, flour	380	360	0.050	50
Cadmium yellow	390	–	–	–
Cellulose triacetate	390	–	0.035	30
Corn flour	390	–	–	–
Cornstarch	390	–	0.040	30
Ethyl hydroxyethyl cellulose	390	–	0.020	30
Ferric ammonium ferrocyanide	390	210	1.500	–
Leather	390	–	–	–
Polyethylene	390	–	0.020	10
Polyvinyl butyral	390	–	0.020	10
Rosin	390	–	0.015	10
Tetranitro carbazole	395	Melts	–	–
Aluminum stearate	400	380	0.015	10
Calcium stearate	400	–	0.025	15

(continued)

Table 4.4 (continued)

Dust	Minimum ignition temperature °C		Minimum explosive concentration	Minimum ignition energy
	Cloud	Layer	g/l	mJ
S-S'-Dimethyl xanthogenethylene bis dithiocarbamate	400	–	0.300	3200
Hydroxyl propyl cellulose	400	–	0.020	30
Malt barley	400	250	0.055	35
p-Nitro-o-anisidene	400	–	–	–
Shellac	400	–	0.020	10
Sodium sorbate	400	140	0.050	30
Anti-pyrin	405	Melts	–	–
Cellulose	410	300	0.045	40
Coffee, instant	410	350	0.280	Did not ignite
Corn dextrin	410	390	0.040	40
Dextrin	410	440	0.050	40
Hexa methylene tetramine	410	–	0.015	10
Hydroxyethyl cellulose	410	–	0.025	40
Hydroxyethyl methyl cellulose	410	–	–	–
Maize starch	410	–	–	–
Melamine formaldehyde resin	410	–	0.02	50
Onion, dehydrated	410	–	0.130	Did not ignite
Para formaldehyde	410	–	0.040	20
Pectin	410	200	0.075	35
Polyacrylamide	410	240	0.040	30
Naphthol yellow	415	395	–	–
Rye flour	415	325	–	–
Antimony	420	330	0.420	1920
Di-t-Butyl-p-cresol	420	–	0.015	20
Dicyclopentadiene dioxide	420	–	0.015	30
Iron, carbonyl	420	230	0.105	100
Peat	420	295	–	–
Polypropylene	420	–	0.020	30
Rayon, viscose	420	–	–	–
Sucrose	420	Melts	0.045	40

(continued)

Table 4.4 (continued)

Dust	Minimum ignition temperature °C		Minimum explosive concentration	Minimum ignition energy
	Cloud	Layer	g/l	mJ
Thiourea	420	Melts	–	–
Tobacco, stem	420	230	Did not ignite	
Walnut shell	420	210	0.035	60
Wheat, grain dust	420	290	–	–
Wood, hard	420	315	–	–
Aluminum-magnesium alloy	430	480	0.020	80
∝,∝' Azo isobutyronitrile	430	350	0.015	25
Caprolactam	430	–	0.07	60
Dehydroacetic acid	430	–	0.030	15
Iron	430	240	–	–
Maize husk	430	–	–	–
Manioc	430	–	–	–
Nitropyridone	430	Melts	0.045	35
Potato starch	430	–	–	–
Sawdust	430	–	–	–
Soap	430	600	0.085	100
Urea formaldehyde resin	430	–	0.02	34
Wheat starch	430	–	0.045	25
Wood, flour	430	–	0.050	20
Almond shell	440	200	0.065	80
Benzotriazole	440	–	0.030	30
Cinnamon	440	230	0.060	30
Cerrovandadium	440	400	1.300	400
Milk	440	–	–	–
Paper	440	270	0.055	60
Polyacetal	440	–	0.035	20
Polymethyl methacrylate	440	–	0.020	15
Rice	440	240	0.050	10
Rubber, crumb	440	–	–	–
Senna	440	–	0.010	105
Sorbic acid	440	460	0.020	15
Wood, soft	440	325	–	–

(continued)

Table 4.4 (continued)

Dust	Minimum ignition temperature °C		Minimum explosive concentration	Minimum ignition energy
	Cloud	Layer	g/l	mJ
2 Acetylamino-5-nitro thiazole	450	450	0.160	40
Acetyl-p-nitro-o-toluidine	450	–	–	–
Bread	450	–	–	–
Coconut	450	280	–	–
Corn cob	450	240	0.045	45
Dinitro stilbene disulfonic acid	450	–	–	–
Ethylene diamine tetra acetic acid	450	–	0.075	50
Ferromanganese	450	290	0.130	80
Lignin	450	–	0.040	20
Lignite	450	200	0.030	30
Milk sugar	450	Melts	–	–
Penta erythritol	450	–	0.030	10
Phenol formaldehyde	450	–	0.015	10
Poly methacrylic acid	450	290	0.045	100
Polyvinylacetate	450	–	0.040	160
Polyvinyl alcohol	450	Melts	–	–
Potato, dried	450	–	–	–
Quillaia bark	450	–	–	–
Steel	450	–	–	–
Wood, bark	450	250	0.020	60
Alfalfa	460	200	0.065	80
Aluminum octoate	460	–	–	–
2 Amino-5-nitrothiazole	460	460	0.075	30
Carboxy methyl cellulose	460	310	0.060	140
Casein	460	–	–	–
Cellulose propionate	460	–	0.025	60
Cellulose tripropionate	460	–	0.025	45
Cork	460	210	0.035	35
3,5-Dinitrobenzoic acid	460	–	0.050	45
Manganese	460	240	0.125	305
Mannitol	460	–	0.065	40
Peanut hull	460	210	0.045	50

(continued)

Table 4.4 (continued)

Dust	Minimum ignition temperature °C		Minimum explosive concentration	Minimum ignition energy
	Cloud	Layer	g/l	mJ
Peat, sphagnum	460	240	0.045	50
Pyrethrum	460	210	0.100	80
Urea formaldehyde molding powder	460	–	0.085	80
Vitamin C	460	280	0.070	60
Polyvinyl pyrolidone	465	Melts	–	–
Rape seed meal	465	–	–	–
Aluminum–lithium alloy	470	400	< 0.1	140
Calcium citrate	470	–	–	–
Coconut shell	470	220	0.035	60
Cotton flock	470	–	0.050	25
Dinitro aniline	470	–	–	–
p-Nitro-o-toluidine	470	–	–	–
m-Nitro-p-toluidine	470	–	–	–
Oilcake meal	470	285	–	–
Starch	470	–	–	–
Crystal violet	475	Melts	–	–
Sodium propionate	479	–	–	–
Lycopodium	480	310	0.025	40
Nitro diphenylamine	480	–	–	–
Piperazine	480	–	–	–
Titanium hydride	480	540	0.070	60
s-Trioxane	480	–	0.143	–
Zinc ethylene dithiocarbamate	480	180	–	–
Coal, brown	485	230	–	–
Fish meal	485	–	–	–
Tobacco	485	290	–	–
Bone metal	490	230	–	–
Dimethyl diphenyl urea	490	–	–	–
Epoxy resin	490	–	0.015	9
Gum tragacanth	490	260	0.040	45
Milk, skimmed	490	200	0.050	50
Sodium zylene sulfonate	490	–	–	–

(continued)

Table 4.4 (continued)

Dust	Minimum ignition temperature °C		Minimum explosive concentration	Minimum ignition energy
	Cloud	Layer	g/l	mJ
Starch, cold water	490	–	–	–
Citrus peel	500	330	0.060	100
Cocoa	500	200	0.065	120
3,5-Dinitrobenzamide	500	Melts	0.040	45
Dinitrotoluamide	500	–	0.050	15
Gum Arabic	500	260	0.060	100
Malcic anhydride	500	Melts	–	–
Nylon	500	430	0.030	20
Polyacrylonitrile	500	460	0.025	20
Polyethylene terephthalate	500	–	0.040	35
Poly isobutyl methacrylate	500	280	0.020	40
Polystyrene	500	500	0.020	15
Sodium 2,2-dichloro propionate	500	–	0.260	220
Tea	500	–	–	–
Vanadium	500	490	0.220	60
Anthracene	505	Melts	–	–
Asphalt	510	500	0.025	25
Carbon, black	510	–	–	–
β-Naphthalene-azo-dimethyl aniline	510	Melts	0.020	50
Polyurethane foam	510	440	0.030	20
Sodium dihydroxy naphthalene disulfonate	510	–	–	–
Sodium thiosulfate	510	330	–	–
Calcium DL Pantothenate	520	–	0.050	80
Carboxy polymethylene	520	–	0.115	640
Cotton linters	520	–	0.50	1920
Fumaric acid	520	–	0.085	35
Gum Karaya	520	240	0.100	180
Isinglass	520	–	–	–
Potassium hydrogen tartrate	520	–	–	–
Sodium secobarbital	520	–	0.100	960
Yeast	520	260	0.050	960
Calcium propionate	530	–	–	–

(continued)

Table 4.4 (continued)

Dust	Minimum ignition temperature °C		Minimum explosive concentration	Minimum ignition energy
	Cloud	Layer	g/l	mJ
Charcoal	530	180	0.140	20
Cottonseed meal	530	200	0.055	80
Dibutyl tin oxide	530	–	–	–
Moss, Irish	530	230	Did not ignite	
Phenol furfural resin	530	–	0.025	10
Sodium toluene sulfonate	530	–	–	–
Calcium silicide	540	540	0.060	150
Chloramine-T	540	150	–	–
2,4-Dichlorophenoxy ethyl benzoate	540	–	0.045	60
Dimethylacridan	540	–	–	–
Monosodium salt of trichloro ethyl phosphate	540	–	–	–
Phenothiazine	540	–	0.030	–
Soya protein	540	–	0.050	60
Tung	540	240	0.070	240
Adipic acid	550	–	0.035	60
Aluminum-iron alloy	550	450	–	–
Aspirin	550	Melts	0.015	16
Calcium gluconate	550	–	–	–
Coumarone-indene resin	550	–	0.015	10
Diamino stilbene disulfonic acid	550	–	–	–
Diazo aminobenzene	550	–	0.015	20
4,4'-Dinitro-sym-diphenyl urea	550	–	0.095	60
Polyurethane foam, fire retardant	550	390	0.025	15
Sodium 2-chloro-5 nitrobenzene sulfonate	550	440	–	–
Sodium monochloracetate	550	–	–	–
Soya flour	550	340	0.060	100
Tellurium	550	340	–	–
Calcium carbide	555	325	–	–
Acetamide	560	–	–	–
Aceto acetanilide	560	–	0.030	20
Acetoacet-p-phenetedide	560	–	0.030	10
Aluminum acetate	560	640	–	–

(continued)

Table 4.4 (continued)

Dust	Minimum ignition temperature °C		Minimum explosive concentration	Minimum ignition energy
	Cloud	Layer	g/l	mJ
P-t-buyl benzoic acid	560	–	0.020	25
Magnesium	560	430	0.030	40
Sodium benzoate	560	680	0.050	80
Bis (2-hydroxy-5-chlorophenyl)-methane	570	–	0.040	60
Cadmium	570	250	–	4000
Dimethyl terephthalate	570	–	0.030	20
Diphenylol propane (Bisphenol-A)	570	–	0.012	11
Hydrazine acid tartrate	570	–	0.175	460
Kelp	570	220	Did not ignite	
Lactalbumin	570	240	0.040	50
Molybdenum disulfide	570	290	–	–
Coal, 43% volatiles	575	180	0.050	50
Naphthalene	575	Melts	–	–
Anthranilic acid	580	–	0.030	35
Chromium	580	400	0.230	140
Dimethyl isophthalate	580	–	0.025	15
Gilsonite	580	500	0.020	25
Sodium amatol	580	Melts	0.140	–
Tea, instant	580	340	Did not ignite	
4-Chloro-2 nitro aniline	590	120	< 0.750	140
4,4'-Diphenyl di sulfonylazide	590	140	0.065	30
Salicylic acid	590	–	0.025	–
Sodium acetate	590	–	0.030	35
Benzoic acid	600	Melts	0.011	12
Coffee, extract	600	–	–	–
Dibutyl tin maleate	600	–	–	–
Dihydrostreptomycin sulfate	600	230	0.520	–
Polymonochlorotrifluoro ethylene	600	720	Did not ignite	
Sodium glucaspaldrate	600	–	–	–
Sodium glucoheptonate	600	–	–	–
Coal, 25% volatiles	605	210	0.120	120
Phthalic anhydride	605	Melts	0.015	15

(continued)

Table 4.4 (continued)

Dust	Minimum ignition temperature °C		Minimum explosive concentration	Minimum ignition energy
	Cloud	Layer	g/l	mJ
Azelaic acid	610	–	0.025	25
Coal, 37% volatiles	610	170	0.055	60
Egg white	610	–	0.14	640
Salicylanilide	610	Melts	0.040	20
Beryllium acetate, basic	620	–	0.080	100
Gelatin, dried	620	480	< 0.5	–
p-Hydroxy benzoic acid	620	–	0.025	40
Monochloracetic acid	620	–	–	–
p-Phenylene diamine	620	–	0.025	30
Diphenyl	630	–	0.015	20
Nigrosine hydrochloride	630	–	–	–
Phthalimide	630	–	0.030	50
Tantalum	630	300	< 0.20	120
Tin	630	430	0.190	80
o-Chloro aceto acetanilide	640	–	0.035	30
Ferrosilicon (45% Si)	640	–	–	–
Aluminum, atomized	650	760	0.045	10
p-Chloroaceto acetanilide	650	–	0.035	20
Chloro amino toluene sulfonic acid	650	–	–	–
p-Chloro o-toluidine hydrochloride	650	–	–	–
Phthalic acid	650	Melts	–	–
Carbon, activated	660	270	0.100	–
Aluminum–silicon alloy	670	–	0.040	60
Anthraquinone	670	–	–	–
Coal, 12% volatiles	670	240	–	–
Coke, petroleum, 13% volatiles	670	–	1.00	–
β-Naphthol	670	–	–	–
Polytetrafluoro ethylene	670	570	Did not ignite	
Polyvinyl chloride	670	–	Did not ignite	
Polyvinylidene chloride	670	–	–	–
Terephthallic acid	680	–	0.050	20
Zinc	680	460	0.500	960

(continued)

Table 4.4 (continued)

Dust	Minimum ignition temperature °C		Minimum explosive concentration	Minimum ignition energy
	Cloud	Layer	g/l	mJ
Saccharin	690	–	–	–
Soot	> 690	535	–	–
Copper	700	–	–	Did not ignite
Isatoic anhydride	700	–	0.035	25
Isophthalic acid	700	–	0.035	25
Phthalodinitrile	> 700	Melts	–	–
Streptomycin sulfate	700	–	–	–
Acetoacet-o-toluidine	710	–	–	–
Pitch	710	–	0.035	20
Polycarbonate	710	–	0.025	25
Molybdenum	720	360	–	–
Boron	730	390	Did not ignite	–
Coal, 8% volatiles	730	–	–	–
Graphite	730	580	–	–
Lampblack	730	–	–	–
Tungsten	730	470	–	–
Coke	> 750	430	–	–
Cobalt	760	370	–	–
Dichlorophene	770	–	–	–
Ferrochromium	790	670	2.00	–
Lead	790	290	–	Did not ignite
1-Methylamino anthraquinone	830	Melts	0.055	50
Tribromosalicyl anilide	880	Melts	–	–
Urea	900	–	Did not ignite	
Beryllium	910	540	–	–
Aluminum-cobalt alloy	950	570	0.180	100
Aluminum-nickel alloy	950	540	0.190	80

References

1. Palmer KN (1973) Dust explosions and fires
2. Eckhoff RK (2019) Method of chemical process safety. 3:7–32

3. Morozzo C (1795) Account of a violent explosion which happened in a flour-warehouse, at Turin, December the 14th, 1785, to which are added some observations on Spontaneous Inflammations. Repertory Arts Manuf 2:416–432
4. Peter Field "Dust Explosion" 1982
5. IEC 60529:2001 Degrees of protection provided by enclosures (IP code)
6. IEC 60034 Parts 5: Degrees of protection provided by the integral design of rotating electrical machines (IP code)—Classification
7. McMillan A (1998) Electrical installation in hazardous areas, 22nd May 1998
8. IEC 60079-0:2017 "Electrical Apparatus for Explosive Gas Atmospheres" Part-0: General requirements
9. Brocon KC, James GJ (1962) Safety in Mine research report 201
10. Vishwakarma RK, Ahirwal B, Singh AK, Kumar A, Sinha a (2007) "Electrical Apparatus for Explosion Prevention" Electrical India, pp 104–108, September 2007
11. Donze' E et al (2015) Minimum Ignition temperature of layer and cloud dust mixtures. J Loss Prevent Process Ind 36:326–334

Chapter 5
Explosion-Proof Motors

5.1 Introduction

Design of explosion-proof motors (Ex-motors), such as flameproof Ex 'd', increased safety Ex 'e', non-sparking Ex 'n', and pressurized motor Ex 'p', is dependent upon the severity of the hazards present in the area. When electrical motors are installed in such areas, there is a risk of explosion. It is not only a matter of safety that determines the placement of electrical motors in hazardous areas, but also a matter of economics and operational efficiency[1]. Now, BIS standards for design evaluation and certification of Ex-motors have been harmonized with those of the International Electro-technical Commission. Hence, the requirements of this standard necessitated certain enhanced design features and special testing requirements before placing these electrical motors in hazardous areas such as petrochemical plants, refineries, underground coal mines. There is a detailed description of the distinct characteristics of flameproof, increased safety, non-sparking and purge-protected motors in this chapter, as well as their testing requirements and testing setup, as well as key design considerations made to meet the stringent demands of hazardous areas. Many factors must be considered when selecting electric equipment for hazardous areas, including the classification of the area (zones 0, 1, or 2), the presence of gases (I, IIA, IIB, or IIC), the temperature class (T1 to T6), and the degree of dust and liquid protection. In order to prevent the possibility of an accident occurring, the IEC specifies special features for use in hazardous areas. Tests must be conducted, and new features must be added to electrical motors in order to prove that they are capable of protecting themselves against explosions.

© The Author(s), under exclusive license to Springer Nature Singapore Pte Ltd. 2023
A. Kumar Singh, *Explosion-Proof Equipment in Hazardous Area*,
https://doi.org/10.1007/978-981-99-2516-2_5

5.2 Flameproof Motors (Ex 'd')

A flameproof motor is used as the basic driving force for the fuel dispensing pump to the hydrogen recycling compressor. In a similar way, a flameproof motor is used in a battery-operated forklift for lifting operations, as well as its backward and forward motion. The motor is an electrical device that converts electrical energy into mechanical work. In areas such as refineries, drilling rings, chemical and gas bottling and handling, flameproof motors are required for operation [2]. Since the formation of an explosive atmosphere cannot be avoided, using flameproof motors is the safest course of action. A mixture of explosive gases, such as hydrogen, acetylene, methane, etc., or combustible dust can cause an explosive atmosphere. This atmosphere can be ignited by an arc, spark, or hot surface. Flameproof electrical motors are intended for use only in zones 1 and 2 [3].

Flameproof motors are designed to produce desired output and are mechanically robust and explosion proof. Terminal boxes and enclosures of flameproof motors are strong enough to endure explosions caused by gas and air mixtures. Throughout the design of the motor, it has been ensured that an explosion inside the motor and combustible product will not be transmitted to the flammable media in the outer atmosphere. The design of flameproof motors is relatively more challenging than the design of increased safety, non-sparking and purge-protected motors. The conditions in underground coal mines are the worst compared to surface area application like petroleum industries [4].

5.2.1 Design Consideration

The following considerations should be taken into account when designing motors:

1. **Safety aspect**: There is a chance of explosion due to the presence of an explosive atmosphere around the operational area of the motor. There is a change of danger of life working around that place.
2. **Restriction in the space**: Due to limitations of space, low height, and narrow galleries, an underground electric motor of compact size should be used.
3. **Adverse environmental condition**: The enclosure of the electric motor should be very sturdy in design to withstand the impact of stones and coal falling onto it.
4. **Severe operating conditions**: Most of the times user or consultant does not know the degree of variation of load at site. The type of resistive load against which the machine will be operating is not well defined. The load may vary considerable depending on site conditions.
5. **Difficulty in maintenance**: Like poor illumination, hot, humid watery, rusty atmosphere, and restricted space in the underground coal mine, is not at all suitable for maintenance.

Following are some considerations for current technology trends:

- The motor should be suitable for use with a variable frequency drive (VFD).
- It should be with high frequency.
- A separate terminal box is provided for the space heater and thermistor.
- In the non-drive end, an insulated bearing is used.
- In frame sizes 80–132, on-line greasing is required.
- Low noise and vibration.
- The temperature class is generally T_5 or T_6.

6. **Cable termination**: It is the most important portion because first customer interface is through the terminal box. Customers prefer big-size terminal boxes, sometimes cable is pre-decided before purchase of motor. Occasionally, the size of the cable is oversize compared to the motor's current requirement.

Cable termination in a flameproof motor is the most important portion. Generally, direct cable entry into the main enclosure of the flameproof motor is not provided. Flameproof motors are comprised of two separate enclosures: the main enclosure and the terminal enclosure. In underground coalmines vibration of the motor is more so the cable should enter in the terminal box through a conduit, double compression, or conical cable gland. Sometimes, conduits have been observed to fracture prematurely in underground coal mines due to increased vibration. For this reason, Group I motors are used with a sealing box and flameproof double compression cable glands.

7. According to IS/IEC 60079-1: 2014 [5], construction features for gas group I/ IIA/IIB motors are different from IIC motors (Fig. 5.1).

Following are the main joints in flameproof electric motor (Fig. 5.2).

Flamepaths and gaps must be in compliance with the code requirements. Different gas groups require different flamepaths and gaps. The gap between the rotating shaft and the stationary bearing cover is greater than that between other joints in which both parts are stationary.

5.2.2 *Test Procedure*

As per IEC 60079-1:2014 [5] flameproof motors are tested by testing scientists/ technical officers from testing laboratories and design engineers from motor manufacturers. There is a need for coordination between the two agencies. As a general rule, motor testing can be divided into two major components:

Heat run tests, performance tests, and hydraulic pressure tests are conducted at the manufacturer's facility. Ingress protection tests, explosion tests, etc., cannot be conducted at the manufacturer's premises. Therefore, these tests are performed at any testing laboratories' premises like CSIR-CIMFR, Dhanbad.

The flameproof motor was submitted for testing in accordance with the instructions provided by the test laboratories. In general, two test holes were made, one

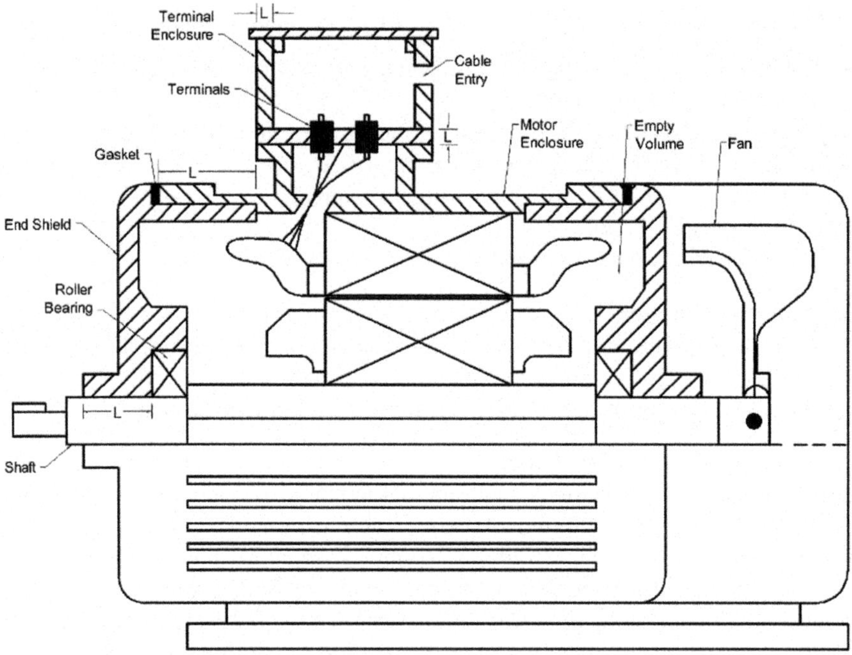

Fig. 5.1 Flameproof motor

for filling gas and the other for connecting either a pressure sensor or a spark plug. Testing laboratories determine the number, size, and location of the holes.

In general, flameproof apparatus is regarded as an enclosure in which the enclosure's strength, gaps, and joint length are adequate to accommodate the explosive mixture present in the atmosphere. A flameproof enclosure must pass a series of tests in order to be classified as flameproof. The tests are the reference explosion pressure test, the overpressure test, and the non-transmission of an internal ignition test [6].

To check the flameproofness of flameproof motors, the following tests are performed in accordance with IEC 60079-0:2017 [7].

5.2.3 Explosion Pressure Test

The purpose of this test is to determine whether the flameproof motor is mechanically sound as it is intended to be used in a combustible media. As part of the testing process, a gas mixture is prepared which is motor enclosure, which is filled inside and ignited by the spark plug. The maximum pressure developed inside the motor enclosure is measured by a piezoelectric pressure transducer. The maximum pressure that develops inside the motor is referred to as the reference pressure. This test is performed five times by changing the position of the pressure sensor and spark plug.

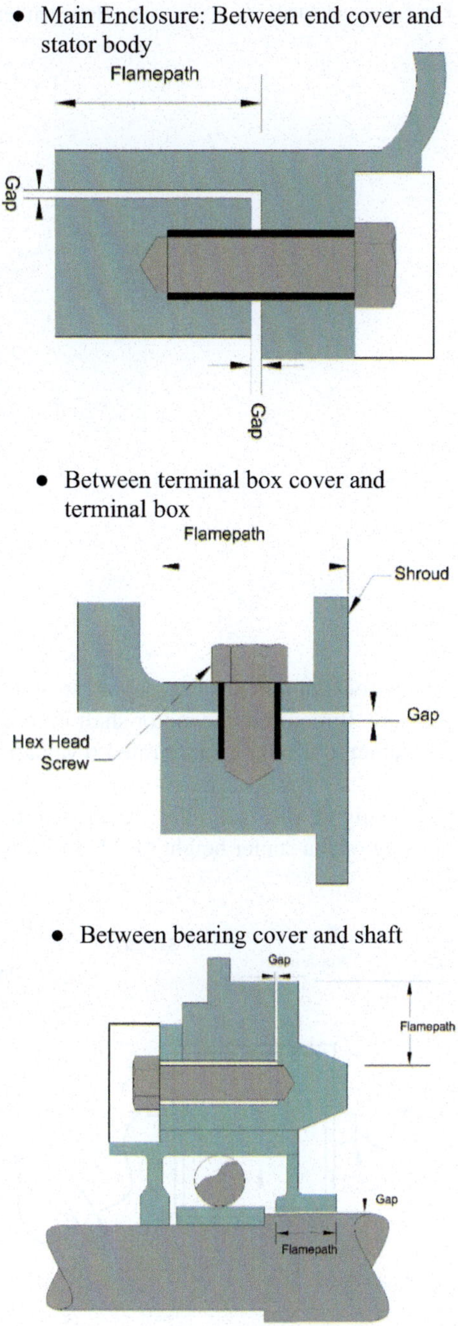

Fig. 5.2 Main joints in flameproof Ex 'd' electric motor (a, b, c, d)

Fig. 5.2 (continued)

- Between terminal stud and terminal box

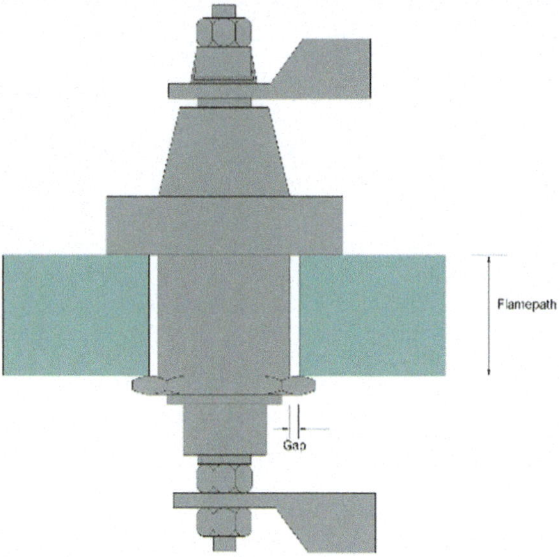

Flameproof motors are tested both in static and rotating conditions for explosion pressure. If required, the motor shaft is coupled with a pony motor which can rotate the flameproof motor under rated revaluation per minute (RPM) [8] (Figs. 5.3 and 5.4).

This figure illustrates the variation in explosion pressure over time for a motor housing with a center height of 90 mm. It was observed that the maximum pressure

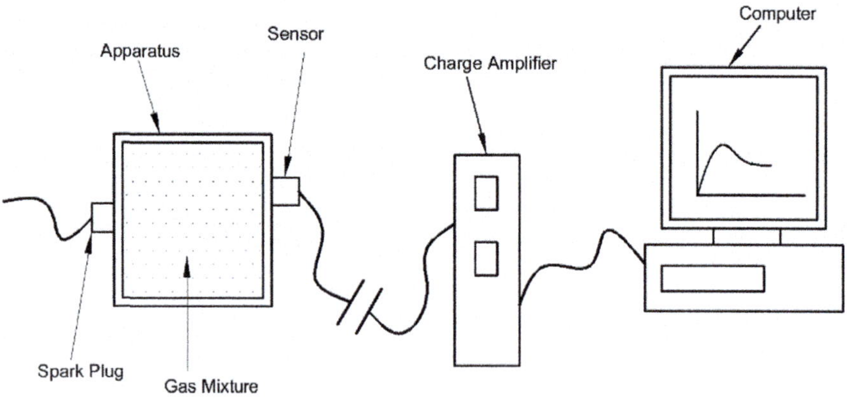

Fig. 5.3 Dynamic explosion pressure recording system

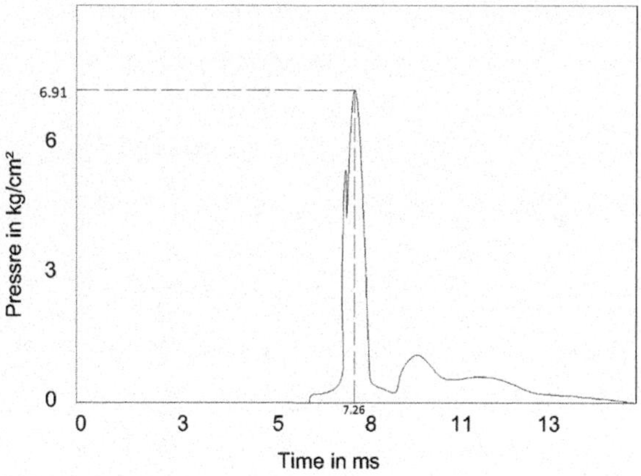

Fig. 5.4 Variation of explosion pressure with respect to time for a motor housing

during the experiment was 6.91 kg/cm^2 at 7.26 ms for the stator housing of a flameproof motor with a center height of 90 mm for a gas mixture of 24% (hydrogen: methane: 85:15) in air.

5.2.4 Overpressure Test

The enclosure should be submitted to static pressure test such that the applied pressure is equal to 1.5 times the reference pressure. For the motor housing, a hydraulic pressure test is conducted at 10.5 kg/cm^2 (approx. 1.5 * 6.91 kg/cm^2) for a minimum of 10 s. No swelling, deformation, damage, or leakage were observed through the motor enclosure's wall or cover.

5.2.5 Non-transmission of an Internal Ignition

Depending on the volume and gas group, each joint of the flameproof apparatus should have a minimum length and maximum gap. This minimum joint length is known as the flamepath, through which the flames release their heat to the surface or body of the apparatus at a temperature lower than the ignition temperature of the gas atmosphere. Flameproof apparatus is placed in an explosion chamber or enclosed in a polythene bag in order to verify this feature. Both the enclosure and the explosion chamber are filled with the same flammable mixture. Spark plugs with low energy are used to ignite the mixture inside the enclosure. If the gas mixture present in the surrounding explosion chamber does not ignite, the test is considered satisfactory. In order to confirm that the outside atmosphere of the equipment is hazardous, the mixture in the chamber is deliberately ignited following the test. A minimum of five tests are conducted. For each test, the mixture is renewed in the enclosure and, if

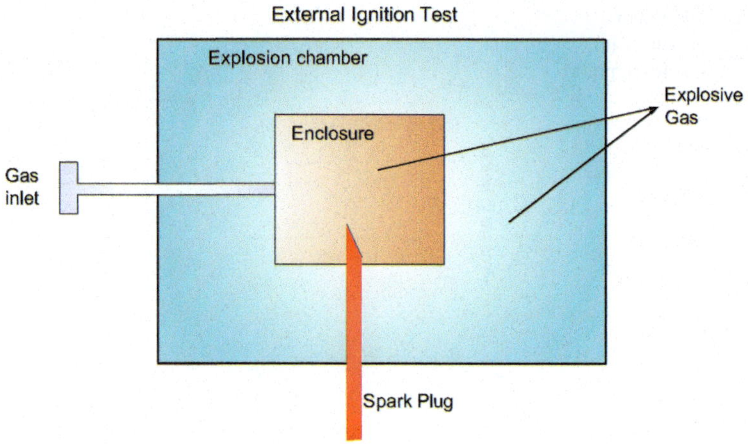

Fig. 5.5 Test setup for non-transmission of internal ignition test

necessary, in the explosion chamber. Hydrogen in the air is 37% of the gas mixture used in this test (Fig. 5.5).

5.2.6 Maximum Surface Temperature Rise Classification

Heat run test of motors

The flameproof motor is tested at full load and at 100% of its rated voltage in order to determine the maximum surface temperature reached during operation. The surface temperature is measured every half hour at DE, NDE, and the motor body in conjunction with the ambient temperature. In the event that the difference between two consecutive temperature readings on the motor body exceeds 1 °C/h compared to the ambient temperature, the motor will be turned off. The surface temperature rise was measured on motor.

The T-class and corresponding maximum temperature is listed in Table 5.1 as per IEC 60079-0:2017.

Stator winding temperature

For flameproof motors, the surface temperature is considered for T-class, but the stator winding temperature must also be considered. The stator winding temperature rise is calculated using the following formula:

$$[(R_H - R_C)/R_C](235 + T_1) + (T_1 - T_2)$$

R_H is the hot resistance measured at the end of the heat run test, R_C is the cold resistance before starting the motor, T_1 is the machine temperature at the time of

Table 5.1 Temperature class and corresponding maximum temperature

T-class	Maximum surface temperature in °C including ambient temperature
T1	> 300
T2	200–300
T3	135–200
T4	100–135
T5	85–100
T6	< 85

cold resistance measurement, and T_2 is the ambient temperature at the end of the examination corresponding to hot resistance.

5.2.7 *Installation of Flameproof Motor*

During the installation of a flameproof motor in a hazardous environment, careful assessment and precaution must be taken. However, the following points should be noted.

- In order to verify the mechanical strength of the motor, a hydraulic pressure test is conducted on its enclosure in accordance with the test certificate.
- The temperature class of flameproof motors as classified by the testing laboratory.
- Deficiency of ingress protection of flameproof motors in accordance with the environmental conditions and the certificate issued by the testing laboratory.

5.3 Increased Safety Motor

Enhanced safety products prevent the explosion of inflammable media by eliminating arcs, sparks, and hot surfaces. The interior temperature of a flameproof enclosure may exceed the ignition temperature of the gas that persists within the enclosure. Nevertheless, safety product concepts aim to reduce unacceptably high temperatures throughout the motor. The reason why generally an asynchronous squirrel cage motor is used as an increased safety motor is due to the fact that it does not generate a spark when operated normally.

Through the design mechanism of the increased safety motor, the surface temperature of the motor is maintained below the incendiary value. Safety motors are manufactured in accordance with code requirements by maintaining clearances, creepage, and insulation. Compared to standard motors, increased safety motors maintain greater clearances and creepage distances by using a high-quality insulating winding system. An increased safety motor is equipped with an overcurrent thermal trip device that disconnects the motor from the line supply within a specified period of time. The increased safety motor provides 100% safety in this situation, however, no protective measures have been implemented for non-sparking motors to address this critical issue. Time t_E is the period of time during which the motor can run with the rotor blocked before it reaches high temperatures. The safety switch for rated current isolates the motor from a line supply within a safe time t_E, thereby preventing the flammable media from being ignited. Because it is lighter and simpler to construct than its flameproof counterpart, it is very popular. The following are some of the special features of motors that have been designed with increased safety in mind.

- Flameproof motors allow ignition within the enclosure of the motor, while motors with increased safety eliminate the possibility of ignition altogether.
- Since increased safety motors are more cost-effective than flameproof motors, they are often installed in hazardous areas. Currently, increased safety motors are only considered for use in zone 2 hazardous areas in India.
- According to the Indian Electricity Rule (2005), increased safety motors may be used in gas generators. The underground coal mines are equipped with suitable monitoring devices for the detection of gases if they exist.

Increased safety motors are characterized by the following features:

1. A high degree of protection must be provided for the enclosure in order to prevent dust, water, or moisture from entering. According to IEC 60529 [9], the minimum degree of protection is IP54.
2. In general, anti-loosening and anti-rotating terminals are used.
3. In accordance with the code requirements, clearance and creepage distances should be maintained.
4. It is recommended that the temperature rise of windings be 10 °C below the temperature rise specified for normal machines or normal motors for an ambient temperature of 40 °C. In contrast, for type 'e' motors the temperature rise is 70 °C by the resistance method.
5. Mechanical clearance between moving parts, such as a fan cover or radial air gap between the stator and rotor, should be in compliance with IEC 60079-7:2015 [10].
6. All gases are classified according to their ignition temperatures into temperature classes T_1 to T_6. It is important that the temperature of the windings and of other parts must not exceed these values if the motor, is running continuously or an install condition, define temperature of the motor should not reach before time t_E second and which should be minimum 5 s. Therefore, the protective switch should be set to operate within time $\mathbf{t_E}$ for relevant temperature class.

5.4 Testing Requirements for Increased Safety Motors as Per IEC 60079-7:2015

5.4.1 Stator

Machines with rated voltages greater than 1 kV must be tested in accordance with clause 6.2.3.1 of IEC 60079-7:2015.

A steady-state ignition test (application of 1.5 times the rated r.m.s line voltage for three minutes in an explosive mixture containing 21% hydrogen in air) and an impulse ignition test are included in 6.2.3.1.

Additionally, if the sum of factors determined by Table 5.2 is more than 6, then anti-condensation heaters shall be employed and the machine shall be constructed to allow special measures to be applied to ensure the machine does not contain an explosive mixture at the time of starting.

Table 5.2 Stator winding discharge risk assessment

Characteristics	Value	Factor
Rated voltage	> 6.6 to 11 kV	4
	> 3.3 to 6.6 kV	2
	1 to 3.3 kV	0
Average starting frequency in service	> 1/h	3
	> 1/day	2
	> 1/week	1
	< 1/week	0
Time between detailed inspections	> 10 years	3
	> 5 to 10 years	2
	> 2 to 5 years	1
	< 2 years	0
Degree of protection	> IP 44	3
	IP 44 and IP 55	2
	IP 55	1
	> IP 55	0
Environmental conditions	Very dirty and wet	4
	Coastal outdoor	3
	Other outdoor	2
	Clean outdoor	1
	Clean and dry indoor	0

5.4.2 Rotor

If the sum total of factors as per Table 5.3 exceeds 5, then the machine shall be tested as per 6.2.3.2 for assessment of possible air gap sparking or the machine shall be constructed to allow special measures to be applied to ensure the machine does not contain an explosive mixture at the time of starting. Testing as per 6.2.3.2 has the following features.

- Testing of the increased safety motor is carried out with stator and rotor.
- Aging process of the rotor cage should be carried out with a minimum of 5 locked rotor tests.
- When the aging process is complete, the increased safety motor is filled with (21 ± 5)% hydrogen in air v/v, and after that, motor should be subjected to 10 locked rotor tests. No explosion shall occur.

Table 5.3 Rotor air gap sparking assessment

Characteristics	Value	Factor
Rotor cage construction	Un-insulated bar fabricated rotor cage	3
	Open slot cast rotor cage ≥ 200 KW/pole	2
	Open slot cast rotor cage < 200 KW/pole	1
	Closed slot cast rotor cage	0
	Insulated bar rotor cage	0
Number of Poles	2 pole	2
	4–8 pole	1
	> 8 pole	0
Rated output	> 500 KW/pole	2
	> 200 Kw to 500 kw per pole	1
	≤ 200 KW per pole	0
Radial cooling ducts in rotor	Yes: > 200 KW/pole	2
	Yes: ≤ 200 KW/pole	1
	No	0
Rotor or stator skew	Yes: > 200 KW/pole	2
	Yes: ≤ 200 KW/pole	0
	No	0
Rotor overhang parts	No compliant	2
	Compliant	0
Limiting temperature	> 200 °C	2
	135 °C < T ≤ 200 °C	1
	≤ 135 °C	0

5.5 Development of Test Facility by BHEL with Consultation of CSIR-CIMFR, Dhanbad, India

Following the implementation of revised standards in January 2008, the industry was facing a very difficult situation. This is because no facilities were available for testing in accordance with the new standards. A virtual halt was brought to the production of non-sparking and increased safety motors as a result of this. At this point, Bharat Heavy Electrical Ltd., Bhopal (BHEL) decided to establish the test facility in Bhopal. Under the guidance of CSIR-CIMFR, scientists, the test setup was developed and tests were conducted. New standards require testing on two fronts, namely:

- Assessment of the potential risk of stator winding discharge.
- A rotor design for possible air gap sparking in first-stage facilities has been developed in order to assess the potential risk of stator winding discharges.

5.5.1 Stator Winding Discharge Risk Assessment

In spite of the precautions taken for increased safety motors, considerable concern has been expressed regarding their use, particularly those that operate at high voltages. This is due to the possibility that potential stator winding discharges could ignite an explosive atmosphere. In Table 5.2, the risk of winding discharge is assessed for stator windings.

If the sum of the total factors exceeds 6, then anti-condensation heaters are employed, and at the time of starting the machine, an explosive gas mixture should not be present in the enclosure.

As part of the testing procedure, the insulation system of the stator winding was also tested pursuant to clauses 6.2.3.1.3 and 6.2.3.1.4 of IEC 60079-7:2015. The following motors were selected for this purpose, with insulation ratings of 6.6 and 11 kV.

- Induction motor with squirrel cage, 3200 KW, 11 kV, 6 poles
- Squirrel cage induction motor with 4600 KW, 6.6 kV, 4 poles.

An explosive gas mixture of (21.5%) hydrogen in the air was immersed in the stator capsule of the machine. The hydrogen concentration was monitored during the test using a flameproof hydrogen analyzer that was calibrated. A stator capsule from a 6.6 kV machine was subjected to a high voltage of 10 kV rms for 3 min, while a stator capsule from an 11 kV machine was subjected to a voltage of 16.5 kV rms for 3 min. The explosion was not observed (Fig. 5.6).

This testing was followed by immersion of the coils of these machines in an explosive gas mixture of $(21 \pm 5)\%$ hydrogen in air. A coil of insulation grade 6.6 kV was subjected to 10 impulses of 16.2 kV peak, 0.2/20-microsecond waveforms. Similarly, a coil of 11 kV insulation grade was subjected to 10 impulses of 27 kV peak voltage. There was no explosion.

(a) (b)

Fig. 5.6 **a**, **b** Testing of stator capsule of 6.6 kV increased safety motor at BHEL Bhopal

Due to successful completion of these tests, increased safety motors can be offered with some provision in the machine to ensure that explosive atmospheres are not present during starting. One such arrangement can be the provision for pre-start purging.

5.6 Rotor Air Gap Sparking Assessment

Based on the factors listed in Table 5.3, the rotor of the machine is assessed for possible air gap sparking.

According to IEC 60079-7:2015, if the sum of the total factors exceeds 6, the machine should be tested in accordance with Clause 6.2.3.2 of the code, and at the time of starting, an explosive atmosphere should not remain inside the enclosure of the motor.

5.6.1 Use of Motor with VFD

Using a motor with a variable frequency drive (VFD) supply causes the winding temperature to rise more than using a basic sine wave supply would. This is a result of additional harmonic losses produced by harmonics present in the drive supply's output. Increase in temperature rise of motors tested with drive is more by 10–15% than motors tested without drive. This observation is based on testing done over wider range of motors and various makes of drive. Also, all consultants specify that the temperature rise of the motor winding is to be restricted to class B limits, even though the motor is wound with class F insulation system.

To meet both conditions, the motor must be rerated. This should be done earlier, before ordering. Nevertheless, authors have found that this point is rarely considered

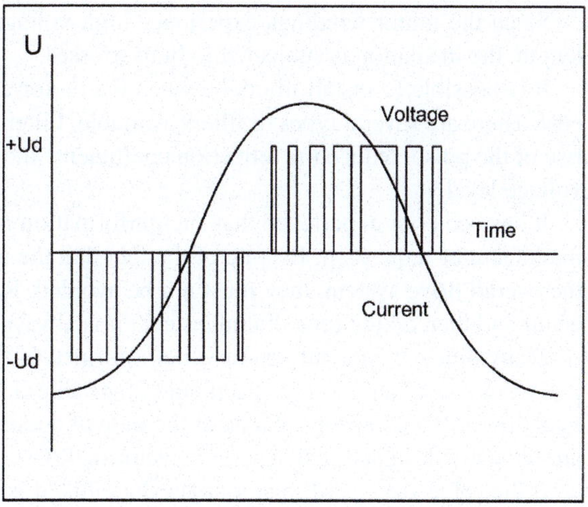

Fig. 5.7 PWM inverter output voltage and current wave front

and that this type of de-ration is rarely considered. Taking into account the above de-ration, we will be faced with a higher initial cost in comparison to the standard frames.

In spite of this, given the importance of the application process, this much additional investment is well worth it as the motor insulation system will be able to last longer (Fig. 5.7).

5.6.2 *Cable Length and Size*

Length of cable used between drive and hazardous area motor is a big issue. Normally, variable frequency drive (VFD) is installed in safe area, and motor is installed in hazardous area. Hence, the length of cable is substantially high. Normally length of cable can be 650 m or more. This much higher length of cable is harmful for health of motor (Fig. 5.8).

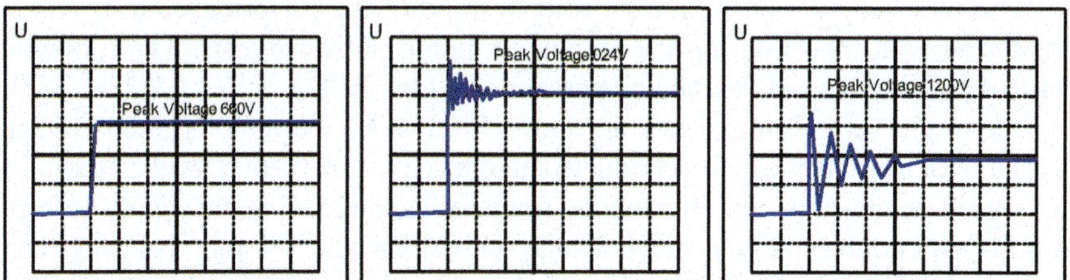

Fig. 5.8 Motor terminal voltage wave forms for varying cable lengths

Since the motor windings experience high-voltage peaks due to the long cable length, the insulation is subjected to high stresses.

It is possible to install filters between the motor and drive at a slight additional cost. There are several types of filters available. Filters such as these limit the rate of rise of the pulse, reduce the reflection coefficient, and consequently reduce the peak voltage level.

It has been our experience that such information is available only when a motor manufacturer requests it. In spite of the fact that the use of filters adds some cost to the overall drive system, they must not be avoided. Rather than allowing failures to occur and then taking remedial measures, it is always better to prevent them.

Many times, the manufacturer is not informed of the cable size when an order is placed. A consultant informs the client of the cable size, which is substantially larger than the current carrying capacity at the time of document approval. Terminal boxes are typically designed with a service factor of 1.5.

If the cable size is still much larger than this, it is necessary to use a flameproof junction box. At this stage, however, the customer is unwilling to use flameproof junction boxes due to additional costs incurred by him and expects the motor manufacturer to provide him with a bigger terminal box to suit his needs. It is possible that the motor manufacturer does not have approvals for cable sizes with such a high service factor.

5.6.3 Combined Testing of the Motor with Drive

In accordance with IEC 60079-10 [11], hazardous area motors intended for use with drives must be tested with the drive and approved by the testing laboratory as a unit. The motor must be tested with the drive every time, even if it has already been tested and approved. This process involves a great deal of coordination between motor manufacturers, equipment manufacturers, consultants, and testing laboratories. In addition, it increases the cost of the project and the length of time required to complete it.

This difficulty can be overcome by recommending some de-ration factors in the standard for motors used in drive supply systems.

With the exception of the locked rotor test, all of the performance tests can be conducted with VFDs during the combined motor and drive tests. The following reasons prevent this test from being performed with a VFD.

Due to the very high locked rotor current drawn by the motor's rated voltage, locked rotor tests are normally conducted at reduced magnetic flux levels. Traditionally, this is achieved by maintaining the frequency and reducing the supply voltage. A VFD may not be able to achieve this, as with VFDs, the V/F ratio is always kept constant, so the magnetic flux level is also kept constant, and this cannot be reduced in a VFD.

The method used to arrive at the t_E value is insignificant, as the magnetic flux level is the same whether it is measured by VFD or direct measurement. The standard t_E

value for utility power supply can already be obtained from the test laboratory during conventional testing.

In addition, the t_E value is mainly used to determine the motor's withstand capability under locked rotor conditions. When using a VFD, there is an inbuilt feature that is available for protection in the event of a locked rotor.

5.6.4 *Separate Terminal Boxes*

Many consultants are asking separate terminal boxes for auxiliary supplied with motors such as space heater thermistor.

Normally, space heaters are required for motors rated 30 KW and higher, whereas thermistors are required for motors rated 75 KW and above. Manufacturers are required to meet the above requirements for all motors with a power rating of 11 kW and above. Auxiliary terminal boxes may be offered, one for a space heater and one for a thermistor.

5.6.5 *High-Efficiency Motors*

Due to the rising cost and shortage of electricity, high-efficiency motors are becoming more and more necessary for use in hazardous environments. The design of these motors differs from that of standard motors. These motors are constructed from more active materials. There are longer cores more copper for winding, and more aluminum for motors. As a result, high-efficiency motors have a different starting performance than standard motors.

In accordance with code IS:12615: 2004 [12], high-efficiency motors require a starting current of 700% with tolerances in accordance with IS 325:1996 [13]. In code IEC 60079-34:2018 [14], selection criteria and motor efficiency are described.

Figure 5.9 shows the constructional features of high-efficiency motor, and Fig. 5.10 shows the cost vs operation hour curve for high-efficiency motor.

1. New fan design
2. Improve steel properties
3. More copper
4. Thinner lamination
5. Reduced air fan
6. Special bearing.

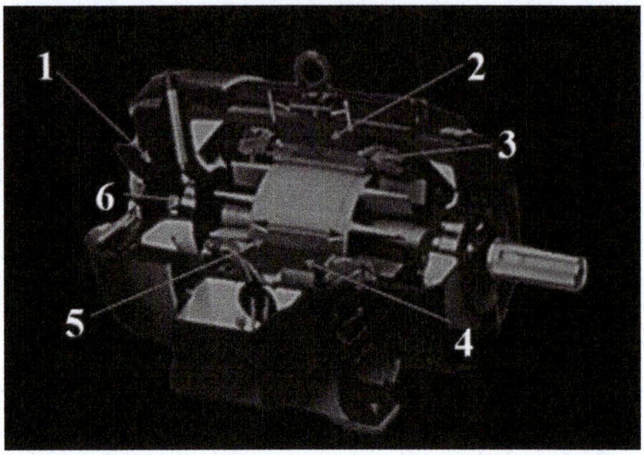

Fig. 5.9 Constructional feature of high-efficiency motor

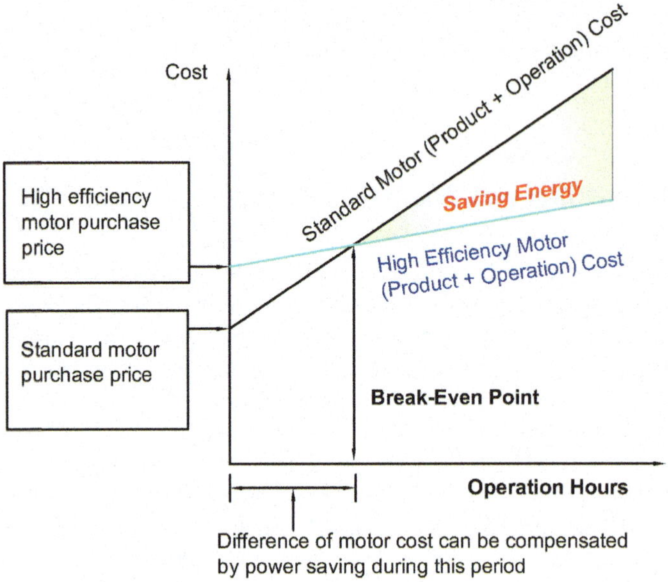

Fig. 5.10 Cost versus operation hour curve for high-efficiency motor

5.7 Non-sparking Motor

There is a great difference between application of non-sparking motor (Ex 'n') and intrinsically safe products (Ex 'i') in hazardous area. Non-sparking product used in zone 2 and intrinsically safe is used in zone 0 and zone 1. Previously, non-sparking types of protection were recognized only in the USA and Canada, but today they are accepted throughout the world. Meanwhile, intrinsically safe product (IS) has been accepted worldwide since its inception and is regarded as the safest form of explosion protection.

However, non-sparking Ex 'n' product is less expensive than intrinsically safe Ex 'i' product. Despite the marginal cost savings, the overall installation is less safe with Ex 'n' than with Ex 'i'.

Non-sparking (Ex 'n') motors are manufactured in the same manner as increased safety motors (Ex 'e'), but do not have the same restriction on lower temperatures as Ex 'e' motors. Due to the lower risk in zone 2 areas, it has the same relaxation in requirements. By restricting the arc, spark, and hot surfaces, the Ex 'n' motor prevents an explosion from occurring. Non-sparking (Ex 'n') machines have some special features that should be noted.

- Terminals that are vibration-proof and non-sparking.
- Clearance and creep age in accordance with IEC 60079-15:2015.
- Rotor bars should be tight in the slot, and brazed end rings should prevent sparking when the motor is started.
- There must be sufficient clearance between the fan and the nearest stationary parts both axially and radially.
- Rotor and stator clearance should be adequate.
- The minimum ingress protection for machine enclosures should be IP 54 for terminal boxes.
- The terminal box should be able to withstand a fault without being damaged.
- The surface temperature of the Ex 'n' motor should not exceed 200 °C.

Normal operation is not subject to surface temperature limitations. It is presumed that the occurrence of flammable media and the starting sequence of the motor does not occur at the same time.

5.8 Harmonization of Indian Standard

Indian Standards have been harmonized based on International Electro-technical Commission (IEC) standards, such as IEC 60079-0:2017 for general requirements, IEC 60079-1:2014 for flameproof motors, IEC 60079-15 for increased safety and non-sparking, and IEC 60079-17 for inspection and maintenance of electrical equipment.

5.9 Purge-Protected Motor Ex 'p'

In this technique, a pressurized panel or motor is purged with clean air or inert gas to prevent ignition. Due to the presence of high-pressure fresh air or inert gases maintained at a pressure of around 5 mg wc relative to the outer atmosphere, flammable gases cannot enter the enclosure. In this way, purge-protected products prevent explosions by preventing flammable media from entering the enclosure. The author

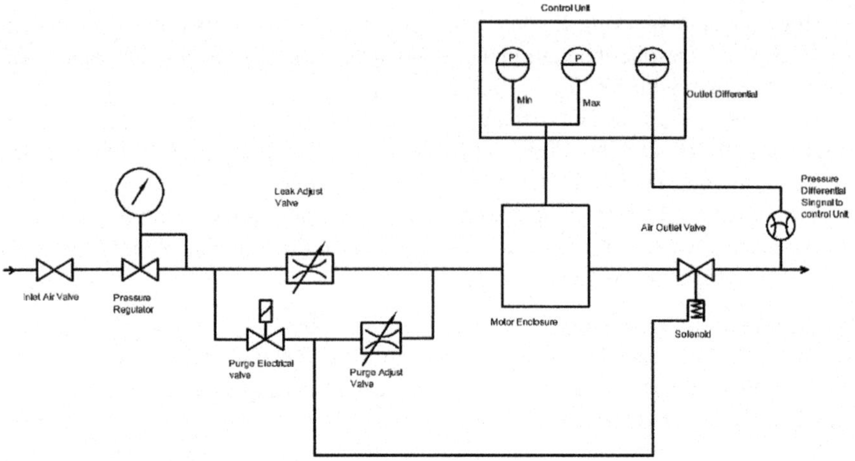

Fig. 5.11 P&I drawing of pressurization control system

discusses different steps involved in the manufacture of a purged and pressurized Ex 'p' protection motor in this chapter.

5.9.1 Pressurized Ex 'p' Motor

First, the pressurized motor must be purged with clean, dry air circulating through the enclosure so that any flammable gas trapped inside will be diluted and forced out. Once the motor has been pressurized, it can be energized. Clean air should compensate for leakage from motor enclosures, and leakage from motor enclosures should be minimized by using enclosures that are as airtight as possible (Fig. 5.11).

Purging sentence correction required

Through the 'Inlet Air Valve', air is allowed to enter the Pressurization Control System, and the inlet pressure can be regulated to 3–5 bars using the 'Pressure Regulation'. After setting the purge time, during pre-start purging, the 'Purge Electrical Valve' Closes and purge air flow through 'Purge adjust Valve' into motor enclosure. The 'Solenoid' opens the 'Air Outlet Valve' with the help of pneumatic signal. If the 'Pressure Different Signal to Control Unit', 'Pmin' and 'Pmax.' values of motor enclosure pressures as sensed by 'Control Unit' are within pre-set values, the purging is started and countdown is displayed in display of 'Control Unit'.

The minimum time required for purging the motor before starting it is:

$$T\,purge = 5 \times V/F$$

where

V = Total free internal volume of the motor including the volume displaced by the motor.
F = Flow rate of the dry air-inert gas.

According to the standard IEC 60079-2:2014 [15], purging and dilution tests must be conducted throughout the test period and the gas concentration at the test points shall be measured or analyzed.

Control System

In the event that the purging cycle is successful, the contacts of the ‘Purge Electrical Valve’ are opened and the pneumatic signal to the outlet ‘Solenoid’ is stopped. As a result, the mechanical ‘Air Outlet Valve’ will close, and thus the ‘Pressure Differential Signal to Control Unit’ will become zero. Pressurized air is introduced into the leakage circuit through the ‘Leakage Adjust valve’ to compensate for leakage from the motor enclosure and to maintain the set value of ‘Pmin’ inside the motor enclosure. As a result of this condition, the motor Ready to Start command is available through the potential-free contacts of the pressurization system.

During purging and pressurization (Leakage Compensation), the ‘Control Unit’ continuously monitors the ‘Pmin’ and ‘Pmax’ values of pressurized air inside the ‘Motor Enclosure’ and gives signal/contact changeover if the set values have been exceeded.

Under various conditions, the behavior of the Pressurization Control System will be as follows:

During Purging

- The purging will not begin if the inlet air pressure (measured at the ‘Pressure Regulator’) is less than 3 bars.
- The purging will not begin if the inlet air flow/discharge is less than 25 lps (the discharge/flow should be measured by the client in the compressed air line). Also, this may be caused by an insufficient opening of the ‘Purge Adjust Valve’.
- Purging will not commence if the ‘Pressure Differential Signal to Control Unit’ is less than the pre-set value of Switch Point Purge Start.
- Purging will not begin if the enclosure pressure during purging is less than ‘Pmin’ or greater than ‘Pmax’.

If any of the above (i to iv) conditions are met, the purging cycle will end without successfully completing. Until the purging cycle is completed successfully and the pressure inside the ‘Motor Enclosure’ is not less than ‘Pmin’ or not greater than ‘Pmax’, the Ready to Start signal will not be available. Once the parameters (i to iv) are within the pre-set desired limits, the purging will automatically restart.

During Pressurization

- In the event that the pressure inside the motor enclosure is less than 'Pmin' or greater than 'Pmax', the Motor Trip signal will be generated at potential-free contacts during pressurization/leakage compensation, and purging will restart automatically for successful completion and to regain the Motor Ready to Start signal.
- It is possible that the pressure drop inside the 'Motor Enclosure' below 'Pmin' is due to a reduced discharge of inlet air from the compressed line of the client (even less than the flow needed for leakage compensation) or heavy leaks from the 'Motor Enclosure', or to the non-availability of pneumatic pressure signals at the 'Control Unit' from the 'motor enclosure' as a result of blocked piping or inadequate opening of the leakage adjustment valve.

Problem during pressurization System

The following are reasons for not starting the purge cycle:

- Unwanted holes in motor enclosures that are not properly plugged result in high leakage.
- There is a problem with the shaft sealing of the motor.
- As a result of inaccuracies in the welding joints of enclosures, leakages occur.
- There is a gap between the base frame and the motor foot that causes air leaks.
- Pipe leaks caused by loose joints.
- Due to the inaccurate selling of pneumatic valves/outlet valves, there is an insufficient differential pressure.
- There is low pressure at the inlet of the control unit.
- Solenoids have a problem.
- There is a problem with the relays and fuses inside the control unit.
- There is a problem with the pressure sensor inside the control unit.

Leakage compensation mode problem

The following are the problems encountered during leakage compensation.

- There may be an overpressure or under pressure in the motor enclosure as a result of an incorrect setting of the leakage valve.
- There is a high level of leakage from pipelines and motor enclosures.
- Cover opening during normal motor operation.
- The minimum pressure point in the motor enclosure has not been selected.
- The minimum pressure point in motor enclosures has not been selected correctly.
- Air flows from the compressed air line.

Solution for the proper pressurization

In order to prevent leakage from the motor enclosure, the following solutions have been proposed.

- Using gaskets to seal the gap between the motor enclosure and cooler.
- Gaskets should be used to seal the covers of all terminal boxes.
- It is necessary to tighten all bolts in terminal boxes, motor frames, heat exchangers, etc.
- The motor should be properly plugged up with all unnecessary holes. Use a sealing material such as M-seal, silicon sealant, etc.
- It is important to properly weld the joint in order to avoid any gaps.
- The labyrinth ring should be used in order to ensure proper shaft sealing.
- The silicone tape should be applied to all joints to ensure tightness.
- It is necessary to clean the pipelines and filter before starting the purge and pressurization unit.
- It is necessary to use an air blower to remove choking and leaks from the instrument pressure sensor tube.
- It is important to select enclosures with the minimum pressure point. When the motor is started and operated, negative pressure is generated at the back of the fan blade. Therefore, it is necessary to select the minimum pressure point accordingly.
- Ensure that the pneumatic valve opening is adjusted properly in order to generate the appropriate differential pressure.
- The proper functioning of all electrical items, such as solenoid valves, relays, fuses, etc., should be checked.
- Check all settings (maximum/minimum pressure, differential pressure, and purging time).

As the Ex ‘p’ motor is used in hazardous areas, it should only be started after successful purging and run with proper leakage compensation/pressurization. It is possible to eliminate all major problems associated with the operation of pressurization systems by considering above solutions.

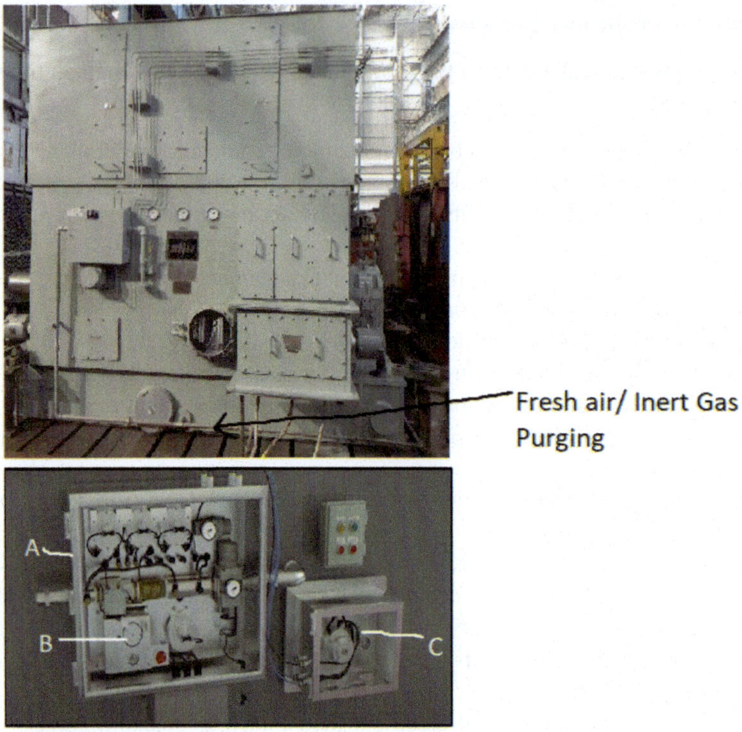

Pressurised Motor

A - Purge Control System

B - Mechanical Timer

C - Vent System

References

1. Arora B, Seth Y, Jena BN (2014) Special features, testing requirements and latest trends in the design of non-sparking, increased safety and flameproof motors. In: 2nd international seminar, on “design, development, testing and certification of ex-equipment” DTEX 2014, Kolkata, India, pp 207–216
2. Vishwakarma RK, Singh AK, Ahirwal B (2007) Safety requirements and selection of electrical equipment for petroleum drilling rigs. In: Proceedings of international conference on “present status and future trend in petroleum industry (PEGJP 07) theme: exploration and processing”, Dec. 2007, pp 6–8
3. IEC 60079-10:2002 Part 14: Classification of hazardous area

4. Desale RS, Bhave D (2009) Hazloc motors—design trends and issues. In: 1st international seminar on "design, development, testing and certification of ex equipment" DTEX 2009, CSIR-CIMFR, Dhanbad, India, pp 73–82
5. IEC 60079-1:2014 "Electrical apparatus for explosive gas atmosphere Part-1: Flameproof enclosure 'd'
6. Singh AK et al (2007) Basic of explosion, construction, testing and certification of flameproof equipment. IEEMA 27:110–114
7. IEC 60079-0:2017. Electrical apparatus for explosive gas atmosphere Part-0: General requirements
8. Vishwakarma RK, Singh AK, Ahirwal B (2010) Explosion pressure development and temperature rise classification of low rating flameproof electric motors. Int J Petrol Sci Technol 4(1–2):1–7
9. IEC 60529 "Degrees of protection provided by enclosures (IP code)"
10. IEC 60079-7:2015 "Electrical apparatus for explosive gas atmosphere" Part 7 increased safety 'e'
11. IEC 60079-10:2020 Explosive atmospheres—Part 10–1: Classification of areas—Explosive gas atmospheres
12. IS 12615: 2011 Energy efficiency Induction Motors
13. IS 325:1996: Three-phase induction motors.
14. ISO/IEC 60079-34:2018—Explosive atmsopheres—Part 34: application of quality management systems for Ex Product manufacture
15. IEC 60079-2:2014—Explosive atmosphere—Part-2: Equipment protection by pressurized enclosure 'p'

Chapter 6
Testing and Certification of Explosion-Proof Equipment

6.1 Introduction

In some areas, such as coal mines and petroleum industries, normal electrical equipment is not permitted. The extraction of coal from underground coal mines, drilling for oil, and refining of oil release flammable gases like methane, hydrogen, acetylene, propane, ethylene, etc., into the atmosphere. When these flammable gases mix with air, they form an explosive atmosphere. In such an explosive environment, explosion-proof electrical equipment is used instead of regular electrical equipment. The use of explosion-proof equipment in hazardous areas is governed by **certain regulations**. In India, the equipment used in underground coal mines is approved by a statutory authority DGMS as per provision of **Regulation 181 (3) of the Coal Mines Regulation 1957 and 75(2) of the Oil Mines Regulation 1984** [1].

Public safety and environmental protection are of paramount importance when dealing with liquid hydrocarbons and flammable gases and vapors. In the process of refining, storing, transporting, or handling petroleum or flammable gases, there is always the possibility of fire or explosion. To eliminate these risks, it is imperative that the associated installations can be designed appropriately and that the necessary equipment can be employed. The Petroleum and Explosives Safety Organization (PESO) in India is responsible for approving equipment installed in petroleum industries. In 1898, PESO was established as the Department of Explosives (DOE) to supervise the manufacture, storage, and transportation of explosives, in accordance with the Explosive Act 1884, substances such as explosives, compressed gases, and petroleum [2]. PESO's major work is to administer the responsibilities delegated under the Explosives Act 1884 and Petroleum Act 1934 and the rules made there under with the motto '**Safety First**'.

The purpose of this chapter is to describe in brief the procedures for obtaining testing and certification and approval of electrical equipment for use in flammable environments such as underground coal mines and surface industries such as petrochemicals, refineries, drilling rigs in India and internationally.

© The Author(s), under exclusive license to Springer Nature Singapore Pte Ltd. 2023
A. Kumar Singh, *Explosion-Proof Equipment in Hazardous Area*,
https://doi.org/10.1007/978-981-99-2516-2_6

6.2 Testing and Certifying Agencies

For electrical equipment to be used in flammable environments, manufacturers must submit an application in the required format to a recognized testing laboratory for testing and certification. The manufacturer first designs their Ex equipment and submits it to the testing authority with the appropriate application and design drawing. The Flameproof and Equipment Safety Laboratory of CSIR-Central Institute of Mining and Fuel Research (CSIR-CIMFR), Dhanbad, is the first recognized testing laboratory in India established in the 1960s for testing and certification of flameproof and intrinsically safe equipment. The laboratory is equipped with modern equipment received from the UK as part of the Indo-UK collaboration program. Other testing laboratories in India include the Electronic Regional Testing Laboratory (ERTL), Kolkata, the Central Power Research Institute (CPRI), Bangalore, Karandikar Laboratories Pvt. Ltd., Mumbai, and Intertek India Pvt. Ltd., Delhi.

6.3 Application

Before any apparatus is sent to Central Institute of Mining and Fuel Research (CIMFR) for test, an application must be made in writing to the Director, Central institute of Mining and Fuel Research, Dhanbad, on the prescribed format Appendix I. The application should be accompanied by a forwarding letter on company's letterhead and three copies of drawings showing flameproof dimension. Appendix II is prescribed format for certification of supplementary, check, and batch testing.

Application for test will be considered when submitted by manufactures of the apparatus or by their duly accredited agents. In exceptional cases, the user of an equipment may also submit application for testing and certification, but such certification will be applicable only for the equipment tested and would not be a type certification covering all the identical items.

A separate application is required for each distinct size or type of apparatus. However, variations in rating if these do not affect the main enclosure and variations in the details of the electrical part as well as alternatives for the mechanical parts or attachments should be included in one application for any particular unit.

If a range of motors is proposed differing in frame diameter with or without variations in core length for each frame size, the whole range may be included in one application.

Manufacturers are required to submit separate applications for testing and certification of intrinsically safe products. This application should be accompanied by a letter on the manufacturer's letterhead. In addition, three copies of drawings are showing the PCB layout, circuit diagram, and entity parameters, i.e., limiting resistance, diode, inductance, capacitance, etc. Application form for certification of intrinsically safe products and PVC/steel cord belt/hoses/brattice sheet is given in Appendixes III and IV, respectively.

The fire and explosion hazard are not only limited to the failure or fault in the electrical equipment used the areas containing explosive gases and vapors, but there may also some chances of fire and explosion due to sources like static charge, frictional incendivity, flame arrester, non-sparking tools, etc.

Following are the non-electrical products, which are tested and certified by this laboratory to service the hazardous area industries.

S. No.	Product	Standard
1	Flame arrester testing	IS 11006;1984 specification for flash back arrestor (flame arrester)
2	Belting test (steel cord, PVC belting) drum friction test flammability test propane burner test electrical resistance test	IS 3181:1992—conveyor belts–fire-resistant conveyor belting for underground mines and such other hazardous application—specification (second revision) IS 1891-1994 (part 1) (Reaffirmed 2000)—conveyor and elevator textile belting—specification (fourth revision) IS 15143:2002—conveyor belting of electromeric and steel cord construction for underground mines and such other hazardous applications—specification
3	Brattice cloth test	IS4595:1969—general requirements for non-sparking tools
4	Non-sparking tools	IS 11884:1986—specification for fire-resistant brattice sheeting made from unsupported plastics IS 4355:1977—specification for fire-resistant brattice cloth
5	Frictional incendivity test for light aluminum alloy	IS 4013:1967—specification for dust-tight electric lighting fittings

6.4 General Criteria

As part of the certification process for flameproof products, drawing certification is mandatory, because technically any testing laboratory can certify the design and drawing, not the product itself. When a disaster occurs in underground coal mines or in the petroleum industries due to a failure of an Ex product, the investigating authority first evaluates the design and drawing that have been certified by the competent laboratory. Preparation of flameproof drawing is a difficult task. Instructions for preparing a flameproof drawing are provided in Appendix V.

For technical reasons, CSIR-CIMFR may allow the manufacturer to withdraw the application before the work has begun, but reserves the right to retain any drawings, technical details, etc. provided by the manufacturer. CSIR-CIMFR has a Customer Facilitation Center/Testing Cell. Testing Cell consists of ten important testing laboratories, including Flameproof and Equipment Safety Laboratory (formerly Flameproof Laboratory). In the first step, manufacturers submit their applications to Director, CSIR-CIMFR, Dhanbad. The Director assigns the application to the Testing Cell. The Testing Cell informs the manufacturer of the test charge based on the required test parameters, the drawing certification, and the number of copies of the original report. After payment of the test charges, the product may be submitted to the Testing Laboratory for testing and certification.

CISR-Central Institute of Mining and Fuel Research, Dhanbad, reserves the right to reject any application that violates any of the following conditions.

1. After receiving the application, the sample should be delivered to CSIR-CIMFR for testing within 270 days.
2. The test charges for the product must be submitted to the CSIR-CIMFR within 90 days of the dispatch of the Price Intimation (PI) by the Testing Cell.
3. Technical data information or clarifications related to the certification of the product must be provided within 90 days.

If an application is considered abandoned, a new application form must be submitted. Although the Director, CSIR, CIMFR, allows reconsideration of abandoned applications, such requests must be made in writing to him with specific reasons for the delay.

A manufacturer or applicant may request to be present during the testing of his product to the Director of the CSIR-CIMFR. The Flameproof and Equipment Safety Laboratory of the CISR-CIMFR will inform the manufacturer or applicant of the date and time of the test in such cases. In the course of testing, applicants may be required to provide technicians, mechanics, or workers to assist in dismantling parts of electrical equipment. The manufacturer may provide special tools for reassembling the part of electrical equipment.

A scientist/technical officer from the Flameproof and Equipment Safety Division reviews the designs and drawings submitted by flameproof manufacturers. Final submission of the correct design and documents in accordance with the code requirements is made to the manufacturer. In the event that there are defects or omissions in the design, drawing, or documents, the same is noted in the test report provided by CSIR-CIMFR to the applicant and the type test/prototype is considered a failure.

The Testing Cell sends the application to the Flameproof and Equipment Safety Laboratory for testing and certification after placing the Catalog number and test charge on the application. Flameproof and Equipment Safety Laboratory attaches an identification number to electrical products, and then, the Head of Department (HOD) of the laboratory marks the application for their scientists/technical officers. Upon receiving the application, these officers prepare a checklist and ask the manufacturer or applicant to provide the required data.

6.5 Test Report and Certificate

In the past, the CSIR-CIMFR issued only test reports to the manufacturer. The test reports contained the details of equipment and test carried out by the laboratory. The conclusion given at the end of the report served the purpose of a certificate. CSIR-CIMFR now issues both the test report and test certificate following the implementation of IEC 60079–1:2014. The reports contain the details of the equipment and tests conducted by CSIR-CIMFR, while the certificate is a concise document describing the equipment and its conformity to a standard set by the International Electrotechnical Commission (IEC). The certificate is an abbreviation for 'Certificate of Conformity' and its International definition is 'A document attesting a product's compliance with a particular standard'.

6.6 Types of Certificate

CSIR-CIMFR, Dhanbad, issues three types of certificates.

1. Prototype/type test.
2. Supplementary report with or without test.
3. Check test.

The original certificate is issued by a competent authority, i.e., a laboratory, following testing and examination of a prototype product. In the event that a manufacturer wishes to make some modifications to the prototype, the laboratory will issue a supplementary report to the manufacturer/applicant.

The laboratory issues two types of supplementary reports:

1. Supplementary with test.
2. Supplementary without test.

In the event that a modification to a flameproof product affects the flameproof ness and mechanical integrity of the prototype product, then a scientist may be required to issue a certificate of modification based on the testing and examination of the prototype product.

In the event that a modification to the flameproof prototype product does not affect the flameproof ness and mechanical integrity of the product, a laboratory is authorized to issue a certificate after examining the design, drawings, and documents.

In some cases, certification is delayed because the manufacturer/applicant insists on a supplementary certificate without testing the modified product. In some cases, original certificates may be issued on the basis of examination of drawings and past data, whereas in some cases supplementary certificates are issued on basis of required testing.

There is also a third type of certificate, which is provided by laboratories, known as a check test. This test is carried out to check the healthiness of electrical product, which is already installed in industries, on the request of end users. In order to perform a check test, the end user sends the electrical product to a testing laboratory. The laboratory carries out the required test. There is no requirement to certify drawings. This test is only conducted by laboratories in order to check the flameproofness of products and their design.

6.6.1 Testing of Prototype/Type Test

Testing of prototypes is intended to achieve the following objectives.

1. A pressure test is performed to determine the mechanical integrity of the product, i.e., to find out whether the enclosure is strong enough to withstand the internal ignition explosion pressure.
2. To find out whether the design of the electrical product can prevent the internal ignition from traveling to the outermost flammable media.

The following are the main tests that are carried out to fulfill the above objectives.

6.6.1.1 Examination of Type Test/Prototype Product

The electrical equipment submitted for testing at the laboratory will be dismantled and the physical dimensions (e.g., flame path, gap, clearance, creepage, etc.) will be compared with the drawing submitted by the manufacturer in order to ensure that the drawing accurately matches the prototype submitted. A review of the product and drawing will be conducted in accordance with International Standard IEC 60079-1:2014 [3]. In the event that the manufacturer/applicant requests that a different standard can be used to test the product, then that standard will be applied. In this case, the manufacturer will provide an English version of the standard to the laboratory. The testing authority may accept or reject the request at its discretion.

6.6.1.2 Determination of Explosion Pressure (Reference Pressure)

The purpose of this test is to verify the mechanical integrity of the flameproof product.

The explosive gas mixture is prepared in a gas chamber in accordance with the code requirements for the particular gas group. It is then filled inside the enclosure and ignited using a spark plug. Piezoelectric pressure transducers are used to measure the maximum pressure developed during explosions. The maximum pressure is referred to as the reference pressure. With a wide measurement range (0–250 bar) and a high sensitivity, the piezoelectric pressure transducer is designed to measure the dynamic explosion pressure that is developed inside the enclosure. Charge amplifiers are used to convert pressure data to voltage, which is then fed to PC-based Data Acquisition

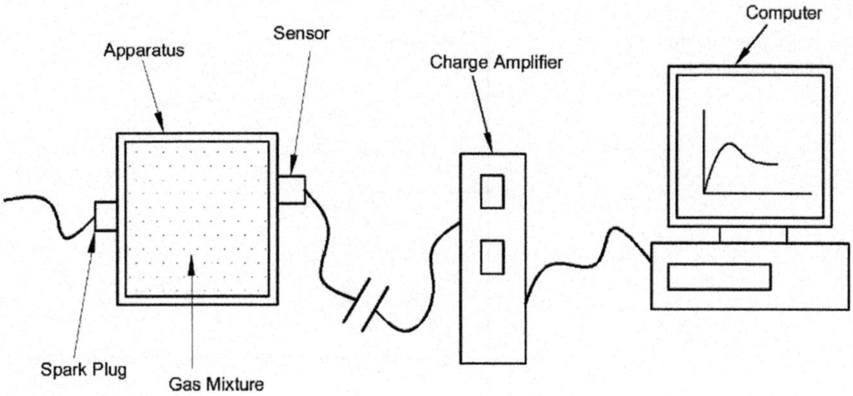

Fig. 6.1 Dynamic explosion pressure recording system

Systems for a graphical display of dynamic explosion pressure (in kg/cm^2) and corresponding time (in m/s) [4] (Fig. 6.1).

It is critical to understand that the reference pressure depends on many factors, including humidity in the air, the orientation of the source of ignition, the energy of the spark, the release of burnt gases through gaps, and also the internal content of the enclosure. Due to the fact that it is unlikely that all parameters will be identical during testing, in order to comply with the standard, the test must be carried out five times, with the highest value being taken as the reference pressure. Spark plugs and pressure sensors are placed at the discretion of the testing authority. The gas mixture is prepared as per the gas group in which the product is to be installed. It is generally recommended that electrical equipment can be tested with all of its internal components. Internal components may be replaced by an equivalent model (Fig. 6.2).

In each test, the maximum pressure developed inside the enclosure is called reference pressure.

The mixture to be used during the test is as follows:

1	Group I enclosure	9.8 ± 0.5% methane in air
2	Group IIA enclosure	4.6 ± 0.3% propane in air
3	Group IIB enclosure	24.0 ± 1% of 85/15 hydrogen/methane
4	Group IIC enclosure	14 ± 1% acetylene in air and also 31 ± 1% hydrogen in air

6.6.1.3 Overpressure Test (Excess Pressure Test)

During the reference pressure test, high pressure is generated for milliseconds; however, during the hydro pressure test, 1.5 times the reference pressure is continuously generated for at least ten seconds, but no longer than sixty seconds. Therefore,

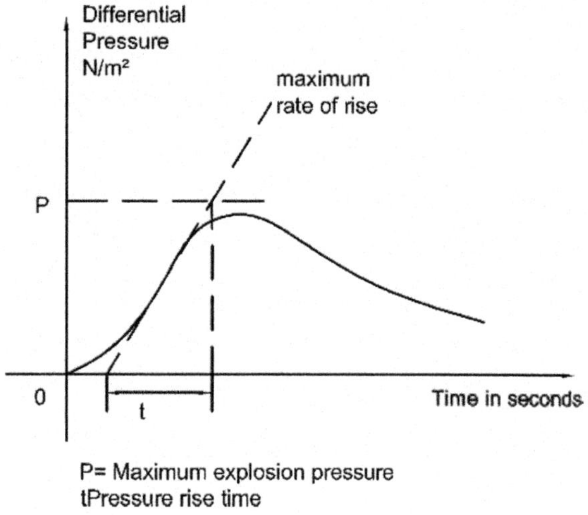

Fig. 6.2 Determination of reference pressure and pressure rise time

the overpressure/hydraulic pressure test plays a crucial role in ensuring that flameproof products are mechanically sound. Nevertheless, it is recommended to carry out hydro pressure at three times the reference pressure if the pressure rise time is less than five milliseconds. As per IEC 60079-1:2014 enclosures with volumes of 2 cm^3 and 100 cm^3, the maximum value of internal overpressure specified is 6 kg/cm^2 for groups IIA, IIB, and IIC. In enclosures with a volume greater than 100 cm^3, the minimum overpressure specified is 8 kg/cm^2 for gas group I and 10 kg/cm^2 for gas groups IIA, IIB, and IIC. For enclosures with a volume smaller than 2 cm^3, an overpressure test is not necessary.

6.6.1.4 External Ignition Test

This test is carried out to check the design of the flameproof products. During external ignition test, flameproof enclosure is placed inside an explosion chamber, i.e., polythene bag and same flammable mixture is filled inside the enclosure as well as the polythene chamber. The flammable mixture inside the enclosure is ignited by a spark plug and generated flame and combustible product during explosion inside enclosure is not communicated in outer atmosphere, then test is considered to be satisfactory. After test mixture inside the polythene chamber is deliberately ignited to check the conformity of the mixture (Fig. 6.3).

The flameproof equipment must be tested at least five times. If necessary, the flameproof equipment may be refilled with the other gas mixture in the gas chamber. As per IEC 60079-1:2014, test mixture for testing of flameproof products is as follows:

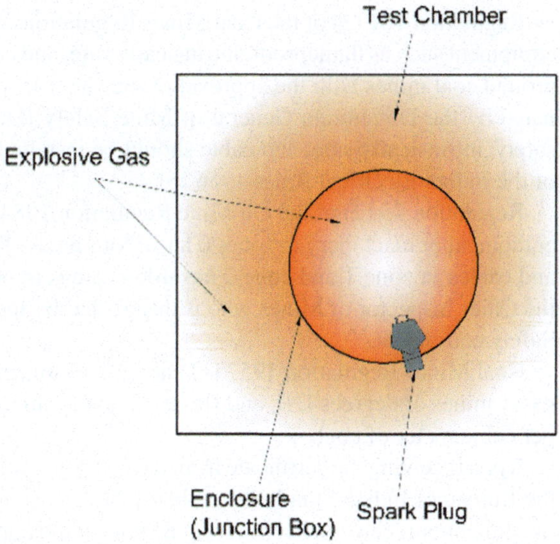

Fig. 6.3 Test setup for external ignition test

1	Group I enclosure	12.5 ± 0.5% (CH_4:H_2::58:42)
2	Group IIA enclosure	4.2% propane or 55 ± 0.5% H_2
3	Group IIB enclosure	37% of hydrogen in air
4	Group IIC enclosure	27.5 ± 1.5% hydrogen in air and also 31 ± 1% acetylene in air

Flameproof products are tested as per IEC 60079-1:2014, Test and Assessment report is prepared as per format provided in Appendix VI, and Type Examination Certificate is issued as per format given in Appendix VIA.

6.7 Approving Authority

The following statutory bodies are involved in the process.

1. Directorate General of Mine Safety (DGMS) for underground coal mines and oil mines, as well as for gassy non-coal mines.
2. The Petroleum and Explosives Safety Organization (PESO), formerly the Chief Controller of Explosives (CCEO) for the petrochemical industry.

Electrical equipment used in underground coal mines

The Mine Act 1952, the Electricity Act 2003, and the rules and regulations framed under these acts govern electrical equipment used in underground coal mines. The Indian Electricity Rules 1956 and the Coal Mines Regulations 1957 [5] govern the installation and maintenance of electrical equipment in mines.

Regulation 181 (3) of the Coal Mines Regulations of 1957 requires that electrical equipment such as flameproof, intrinsically safe, and cable can only be used in underground coal mines with the approval of the Chief Inspector of Mines, who is designated by the Directorate General of Mine Safety (DGMS). Flameproof, increased safety, intrinsically safe, and cable should be installed in accordance with Rule 126 of the Indian Electrical Rules 1956 [6].

Regulation 75 (2) of the Oil Mine Regulation 1984 prohibits the use of electrical equipment or machinery in zones 0 hazardous areas. The use of electrical equipment and cables in zone 1 and zone 2 hazardous areas in oil mines must be approved by the Chief Inspector of Mines who is designated by the Directorate General of Mine Safety equipment.

Coal Mines Regulation 1957 [5] categorizes underground coal mines in India as gassy mines of degrees I, II, and III based on the amount of inflammable gas present per cubic meter of coal.

Upon receiving the certificate from the test house, the manufacturer should contact the Bureau of Indian Standards to obtain the license for the product. As soon as the manufacturer receives a license from BIS, it should approach the DGMS for approval of gas group I (underground coal mines and oil mines). It is the policy of DGMS to require field trials of the product for three months prior to granting approval for suitable use in underground coal mines and oil mines.

During the field trial, the performance of electrical equipment is checked by Mine Management. Electrical equipment is approved by the Directorate General of Mines Safety upon receipt of a satisfactory performance evaluation report from both the end users and an officer of the Directorate General of Mines Safety. A change in the design of electrical equipment suggested by the end user or an officer of the Directorate General of Mines Safety must be carried out by the applicant/manufacturer before approval is granted.

After getting the certificate from competent laboratory, manufacturer applies their product approval to DGMS for use in mines a prescribed format given in Appendix VIII.

In consultation with the Chief Engineers of coal and oil companies, the Directorate General of Mine Safety has selected coal mines for field trials. The DGMS approved the use of electrical equipment in underground coal mines based on the satisfactory results of the field trial.

The Petroleum and Explosives Safety Organization (PESO) has separate procedures for approving indigenous and foreign equipment. Following are the broad outlines of the procedure.

Indigenous equipment

A review of the following documentation is conducted in order to assess the suitability of the equipment.

1. The prototype test report issued by the testing laboratory should validate the design of the equipment in accordance with the applicable standards.

2. The infrastructure capability of a manufacturer is evaluated by a competent authority, such as the Bureau of Indian Standards (BIS). BIS is also responsible for ensuring the quality of products. As part of the quality surveillance scheme, BIS randomly selects products from the manufacturer's shop and sends them to a test house for cross-checking the flameproof product design. Afterward, BIS grants the product a license.

Foreign equipment

Testing procedure in different countries

The type of tests to which a flameproof enclosure is subjected in various countries is broadly similar, but there are variations in details. There are two groups of tests. The first group consisting of reference pressure and overpressure tests is necessary to ascertain the strength of the enclosure, whereas the second group of test consisting of non-transmission test is to ascertain the flameproofness of the enclosure.

The differences in some aspects of testing as prevalent the different countries have been mentioned below. (The list is however not exhaustive).

There is minor difference in composition of gas mixture used during reference pressure test and flameproofness (non-transmission) test [7] (Table 6.1).

Table 6.1 Minimum specified pressure during overpressure test in various national standards—Group I enclosures

Minimum value of specified pressure kg/cm^2	Volume of enclosures			Remarks
	Below 2 cm^3	2–100 cm^3	Above 100 cm^3	
India		6	8	
Japan JIS 00901-1978	Inherent strength considered adequate	6	8	In case of pressure piling 1.5 times the pressure actually measured
Germany VDE 0170-1970		6	8	
U.S.A 2C - 1968		2 times the reference pressure. For casting or welding construction 10.6 kg/cm^2		
U.K. BS: 4683-1971	3.52 kg/cm^2			
CENELEC	3.52 kg/cm^2. If reference pressure cannot be measured and dynamic method not possible 10.5 kg/cm^2			

6.8 ISI Certification Mark

Product certification by the CSIR-CIMFR and approval by statutory authorities are performed on prototypes provided by the applicant/manufacturer, who ensures that the rest of the items produced are wholly in accordance with the prototypes certified and approved. In order to maintain quality control, Bureau of Indian Standards (BIS), a competent authority, conducts shop surveillance of manufacturers.

6.9 Time Factor Involved in Testing, Certification, and Approval

The flameproof manufacturer submits their application along with drawings to the Director, of CSIR-CIMFR, Dhanbad, for testing and certification.

The Director forwards the application to the Flameproof and Equipment Safety Laboratory of CSIR-CIMFR.

A Scientist/Technical Officer from this laboratory submits this application to the Testing Cell for communication of the testing charges to the concerned manufacturer.

Upon receiving the test charge from Testing Cell, that application is forwarded to Test Laboratory for testing.

The initial phase of the process is for the Scientists and Technical Officers to prepare the checklist and to thoroughly scrutinize the flameproof drawing.

Any missing data are communicated to the concerned manufacturer/applicant to provide the information to CSIR-CIMFR. In order to ensure that the required testing of the product is completed on time,

The following observations should be noted.

- Testing and issuing the report took only a minor fraction of the total time taken by CSIR-CIMFR.
- Approximately 80% of test reports are dispatched within three months, which is significantly less than foreign laboratories in the European Union and the USA.

In the case of indigenous equipment, delays can occur for the following reasons:

- The application form contains incomplete information regarding equipment.
- The flameproof drawing is incomplete.
- Failure to submit the equipment and test charges on time.
- Manufacturers are unnecessarily requesting supplementary reports without test certificates when technically this is not feasible.
- Factors other than those controlled by CSIR-CIMFR.

6.10 For Avoiding Delay

CSIR-CIMFR, Dhanbad, has developed a flowchart of certification process for the quick testing and certification of flameproof equipment, as shown in Fig. 6.4. The earlier discussion and flowchart indicate that the manufacturer has a role to play in expediting the testing process at the CSIR-CIMFR. It is the manufacturer's responsibility to submit a proper application and drawing, as well as test charges and a prototype product as soon as possible. In the event that incomplete documents are submitted to CSIR-CIMFR, the case will be classified as pending, i.e.,

1. Applications that are complete.
2. CSIR-CIMFR guidelines for three sets of drawings.
3. Fees associated with testing.
4. Prototype product for testing.

The four items have been submitted in time as shown in the flowchart Fig. 6.4 and test holes have been drilled in the prototype in consultation with the testing authority. This will make it easier to obtain the CSIR-CIMFR certification. For light fittings of different ratings, a different rating bulb and spare glasses should be provided to the

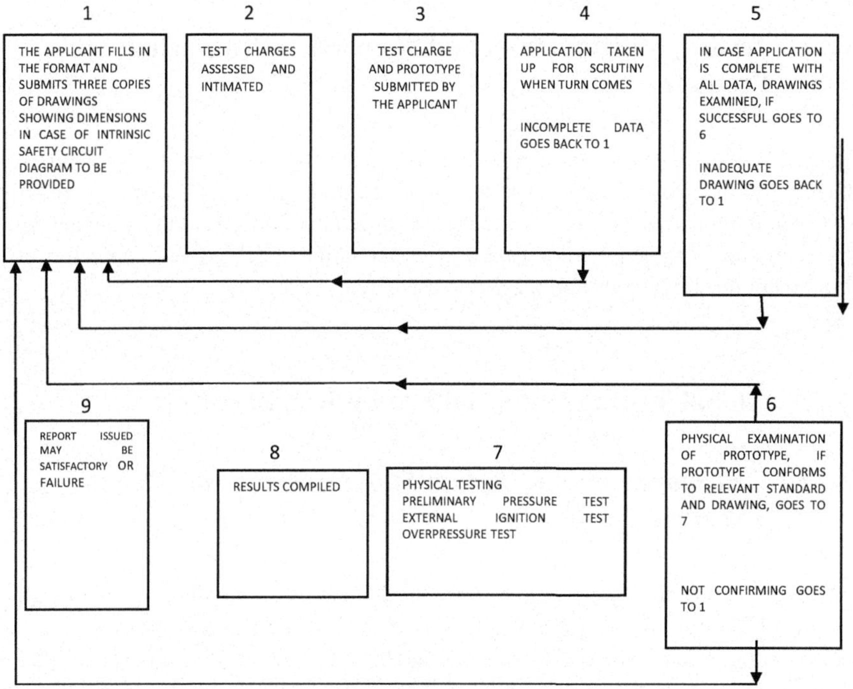

Fig. 6.4 Flowchart of certification procedure in CSIR-CIMFR, Dhanbad, India

Table 6.2 Important test houses in European Union and American continent

Laboratory		Address
FM	Factor Mutual Research	1151 Boston Providence Turnpike, Norwood, Massachusets 02,062, USA
BASEEFA	British approvals' service for Electrical Equipment in Flammable Atmosphere	Harpur Hill, Buxton, Derbyshire, SK 17 9JN
UL	Underwriters Laboratories, Inc	207 E Ohio Street, Chicago, Illinois 60611
HSE (M)	Mining Apparatus	Certification Support Unit, Harpur Hill, Buxton, Derbyshire, SK 17 7JN
PTB	Physikalisch Technische Bundesanstalt	33 Braunschweig Bundesallee 100 Germany
CSA	Canadian Standard Association	Toronto Corporate Head Office, 178 Rexdale Blvd. Toronto, Canada

laboratory. The applicant/manufacturer should take prompt action to implement the above-mentioned lines, which will reduce the testing time significantly.

6.11 Testing, Certification, and Approval of Imported Product

Several types of electrical equipment are imported from the UK, the USA, France, Germany, and Poland. The products used in underground coal mines and the petroleum industries have been tested and certified by their national testing agencies in accordance with their national applicable standards. In Table 6.2, some of the most prominent test houses around the world are listed.

6.12 Global Testing and Certification Requirement

6.12.1 Harmonization of Indian Standards and IECEx Scheme

The Bureau of Indian Standards has been harmonized with the International Electrotechnical Commission (IEC) since 2004. Before 2004, there was a substantial difference among BIS, CENELEC, UL, PTB, and CSA in terms of technical standardization. Due to these regional differences, technical barriers (non-tariff barriers) resulted in repetitive re-testing and re-certification of electrical equipment when moving from one nation to another.

For example:

1. Motors used in hazardous areas in Russian Federation, Kazakhstan, or Belarus must be marked with the EAC (Eur Asian Conformity Mark), formerly GOST-R. In order to accomplish this, a registered notified body of the Customs Union (CU) must conduct a product conformity assessment and a quality assurance conformity assessment in accordance with the EAC codes and standards.
2. The placement of Ex motors in flammable media in Brazil requires certification from INMETRO, its national accreditation body.

IECEx (International Electro-technical Commission System for Certification to Standards Relating to Equipment for Use in Explosive Atmospheres) was developed to eliminate these technical barriers to global free trade. The main objective of this scheme is to facilitate international trade by eliminating the need for multiple national certifications. Table 6.3 shows the list of the countries and their administrations that are members of IECEx.

The Bureau of Indian Standards represents India in the IECEx system. The purpose of a unified certification scheme is to harmonize all the local and regional codes into one standard. Harmonization has resulted in the elimination of the differences between the IEC and NEC. India and the rest of the world use IEC classifications, with the exception of the American continent. The National Electrical Code (NEC) classification system is used in the USA.

The following are some examples of harmonization processes that have already been established.

- Currently, all European standards applicable to hazardous areas have been harmonized with the IEC. They are called EN/IEC instead of EN.
- Standards in South Africa are now harmonized with IEC standards and referred to as SANS60079.
- For class 1 hazards, the NEC has accepted the IEC zone classification system.
- The Canadian Electric Code (CEC) has changed the definition of Class 1 to three zones.
- All of the member countries of the IECEx system are making tangible efforts or have already harmonized themselves with the IECEx system.

However, the world is still divided. As a matter of fact, most countries still maintain an administrative body to grant permission for installation after certification.

For instance,

- PESO in India accepts both IECEx and ATEX certifications for electrical equipment. However, manufacturers still have to follow the entire application process for obtaining permission for installation from PESO.
- USA's default installation is primarily based on the 'Division system'. Furthermore, OSHA has not permitted the use of IECEx reports for the certification of equipment to be used in Division areas. Therefore, the trade barrier still exists.

Table 6.3 IECEx member countries and its administration

IECEx member countries	Administration
India (IN)	Bureau of Indian Standards
UK (GB)	British Electro-technical Committee
Slovenia (SI)	Slovenian Institute of Quality and Metrology
Romania (RO)	Insemex Petrosani Federal Agency on Technical
Switzerland (CH)	Electrosulsse
Germany (DE)	Deutsches Komitee der IEC
Malaysia (MY)	Department of Standards Malaysia (Standards Malaysia)
Netherlands (NL)	Netherlands National Committee of the IEC
Turkey (TR)	Turkish Standards Institution (TSE)
Poland (PL)	Urzard Dozoru Technicznego (UDT)
New Zealand (NZ)	Standards New Zealand
Singapore (SG)	Spring Singapore
USA (US)	US National Committee of the IECEx
Italy (IT)	CEI-Comitato Elettrotecnico Italiano
Australia (AU)	Standards Australia
Korea (KR)	Korean Agency for Technology and Standards KATS
Canada (CA)	Canadian National Committee (CNC/IEC)
Brazil (BR)	COBEI—Comite Brasilerior De Electricidade, Eletronica, Iluminacaoe Telecomunicacoes
France (FR)	Laboratories Central des Industries Electriques LCIE
South Africa (ZA)	South African Flameproof Association
Russia (RU)	Regulating and Metrology (GOST)
Hungary (HU)	Hungarian Standard Institution
China (CH)	Certification and Accreditation Administration of the People's Republic of China
Sweden (SE)	SEK
Denmark (DK)	Dansk Standard
Norway (NO)	Norsk Elektroteknisk Komite (NEK)
Czech Republic (CZ)	Physical technical Testing Institute
Japan (JP)	Japanese Industrial Standards Committee (JSIC)
Croatia (HR)	Ex Agency (Agency for explosive atmospheres)
Finland (FI)	SESKO Standardization in Finland

Nevertheless, reservations are declining at the government and administration levels, which are a positive development. As statutory bodies lift trade barriers, exporting to foreign countries will be easier.

For example

- Now the US Coast Guard accepts the Zone System and IECEx.
- In addition to national certification, Australia, New Zealand, Singapore, and India have incorporated IECEx into their national legal requirements.
- The Customs Union (Belarus, Kazakhstan, and the Russian Federation) and Brazil, which have their own accreditation bodies, are willing to accept IECEx.
- IECEx has been endorsed by the United Nations Economic Commission for Europe (UNECE).

In conclusion, although there are some challenges in achieving the goal of a unified certification scheme, IECEx certification can be viewed as a passport allowing companies to export their products internationally.

6.12.2 IECEx Certification Scheme

Under the IECEx scheme, electrical equipment is tested in accordance with the International Electro-technical Commission (IEC) standard series IEC 60079 for use in hazardous areas. The products tested and certified by any IECEx TL are acceptable worldwide.

The IECEx system covers the following.

- The certified electrical equipment under the IECEx scheme used in explosive atmospheres (IECEx 02 scheme).
- IECEx certified service facilities scheme covering services such as repair and overhaul of Ex equipment (IECEx 03 scheme).
- IECEx conformity mark licensing system (IECEx 04 Scheme).
- IECEx scheme for certification of personal competence for explosive atmosphere (IECEx 05).

6.12.3 Objective of IECEx Scheme

The aim of this scheme is to facilitate world trade in electrical equipment and services intended for use in explosive environments in the following ways.

- Reduce the testing and certification costs of manufacturers while maintaining an appropriate level of safety.
- Reduce barriers to entry into emerging markets.
- Reduce the time to market.
- Deliver explosion-proof-related services across international borders in a timely manner.
- Enhance international confidence in the assessment process of Ex equipment, services, and personnel.
- Provide a global database of issue reports, certificates, and licenses.

Ultimately, the IECEx system and its schemes are intended to achieve the acceptance of **'one standard, one certificate, and one mark'** throughout the world.

6.13 IECEx Product Certification

6.13.1 Flowchart for IECEx Certification

- Design and manufacturing of Ex Component or Equipment by manufacturer as per IEC standards.
- Preparation of certification drawing and instruction manual.
- Submission by manufacturer of prototype along with application, certification drawing, and instruction manual to Ex CB for certification.
- Testing by Ex TL associated with Ex CB.
- Preparation of Draft Ex TR and Draft IECEx Conformity of Certificate (CoC) after successful testing and submission to manufacturer.
- Finalization of Draft Ex TR and Draft IECEx Conformity of Certificate (CoC) by Ex TL as mutually agreed with manufacturer.
- Review of Ex TR and IECEx CoC by reviewer of Ex TL.
- Submission of Ex TR and IECEx CoC for approval by Ex CB.
- Publication by Ex CB of IECEx CoC on IECEx web after successful approval by Ex CB.
- Issue by Ex CB of soft-signed copy of IECEx CoC, Ex TR, and stamped drawing to manufacturer (Fig. 6.5).

Note: IECEx CoC can be issued subject to manufacturer has valid QAR for applicable type of protection. Refer flowchart for QAR, which can happen before or in parallel to testing by same Ex CB or a different Ex CB.

6.13.2 Flowchart for IECEx Quality Assurance Record (QAR)

- Manufacturer should be an ISO 9000 certified company.
- Manufacturer implements ISO IEC 80079-34 for manufacturing of Ex products of desired type of protection as per IEC.
- Manufacturer applies for QAR to ExCB along with Quality Manual for applicable type of protection
- ExCB conducts audit of manufacturers' factory as per ISO IEC 80079-34 for applied type of protection.

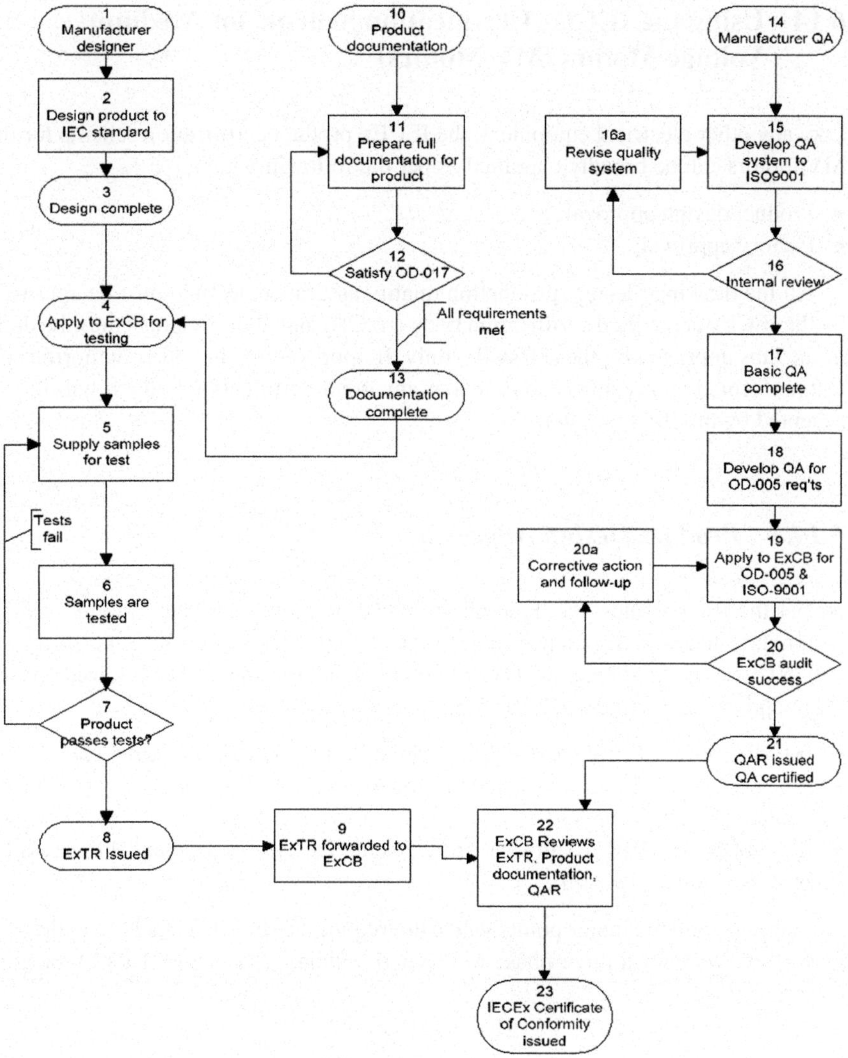

Fig. 6.5 Flowchart of certification of Ex product as per IECEx

- Publication by ExCB of IECEx QAR for particular type of protection on IECEx web after successful audit and closure of NCs, if any.
- Issue of soft signed copy of IECEx QAR to Manufacturer by ExCB.

6.14 Using the IECEx Certification Scheme for Medium Voltage Motors (MV Motors)

Like any other electrical equipment, the IECEx product certification scheme for the MV motors can be broadly classified as per the following.

- Product design approval.
- Factory approval.

All the drawing, design, production quality assurance, testing must be approved by the IECEx accredited certification body (ExCB) and all the testings must be done in the labs approved by these ExCBs, only. If approved by any other authorities or, test performed in any other lab or, under any other certification body, it will not be accepted by the IECEx system.

6.14.1 Product Design Approval

As per the IECEx rules, the three-phase induction motor shall be designed, manufactured, tested, and dispatched in accordance with harmonized IEC standard, i.e., IEC 6079 series of standards. The following standards should be followed for the 'Flameproof', increased safety, and non-sparking motors.

- IEC 60079-0:2017: Explosive Atmosphere Part 0: General Requirements.
- IEC 60079-1:2014: Explosive Atmospheres Part 1: Equipment Protection by Flameproof Enclosures 'd'.
- IEC 60079-15:2010: Explosive Atmospheres Part 15: Equipment Protection by Increased Safety Ex 'e/n'.

Following are the major points which are required to be taken care of, equipment protection-wise, for a three-phase MV (rated voltage greater than 1 kV) induction motors.

6.14.1.1 Flameproof Motors (Ex 'd')

Motor flameproof features (such as flame paths, gaps.) must comply with the gas group requirements and the gross enclosure volume specified in the test report. Flame paths between the shaft of a rotating machine and the gland are very important, and they may be constructed differently with motor driving ends (DE) and non-driving ends (NDE). As a result of non-uniform radial clearances, sleeve bearings should not be used in gas group IIC for motors. The rolling bearing should be replaced with an identical bearing, and if necessary, the permissible radial clearance between the bearing and the gland should be checked. It is important to maintain clearance between the internal fan and its hood.

6.14.1.2 Maintenance of Increased Safety Ex 'e' and Non-sparking Ex 'n' Equipment

Ex 'e' and Ex 'n' equipment are generally designed to provide protection against excessive temperatures, arcs/sparks, weather, dust, and impacts. The cover should only be opened for a brief period of time and the equipment shouldn't be left unsupervised without protection. In the event that dust or moisture has accumulated, it should be removed. In order to ensure adequate current-carrying capacity, the terminals must be sized appropriately and have enough cross-sectional area. Clamping should also be vibration-proof and anti-loosening. There is a minimum clearance and creep age distance that must be maintained between conducting parts. The temperature rating of the motor/junction box must not be exceeded at any time during normal operation, or under any specified condition to prevent the ignition of an explosive atmosphere. To ensure that the limiting temperature cannot be exceeded during normal service, the windings must be protected. This means that if the motor suffers from a locked rotor situation, it must be disconnected from the supply within tE time frame. The protective relay must be checked according to the maintenance schedule.

6.14.1.3 Maintenance of Pressurized Equipment Ex 'p'

In the case of pressurized equipment, keeping a specified minimum overpressure within the enclosure is important because Ex equipment's safety depends on it. About 2 mbar or 1.5 times of maximum pressure should be sustained by the enclosure whichever is greater. During the maintenance, it should be remembered that Ex 'p' equipment has a partial or fully purged system. Non-incentive material should be used to manufacture the cover of the enclosure. The entry point of the cable should be sealed with a sealing compound. The inlet and outlet tubes are used in the purging system such that an overpressure of 5mmWc should be maintained. Special fasteners are used for doors and covers for the pressurized unit. The proper sealing compound should be applied to maintain the ingress protection and minimize the protective gas leakage through the pressurized enclosure. All audio alarms/annunciators should be checked properly. The pressure level gauge for low-/high-level protection should be checked properly. Check electrical and mechanical door interlocks in Ex 'p' panel. The pressure gauge, flow meters, time delay relay, etc. should be calibrated and deviation should be considered at the time of purging or pressuring of the enclosure.

- Flamepath and 'Flameproof' joints at different parts of the enclosure and its joints.
- Requirements of the flame erosion tests, if cemented joints are not used.
- Determination of reference explosion pressure by actually creating an explosion inside the enclosure.
- Hydraulic overpressure test at 1.5 times the reference explosion pressure.
- Thickness of the enclosure as well as the part of the enclosure suitable for the explosion pressure.
- Flame erosion test for the terminal blocks of not having a cemented joint.

- Non-metallic material (on which the type of protection depends) selection, specification, and thermal endurance test as per IEC 60079-0:2011, Cl. No. 7.
- Selection of fasteners and holes for fasteners as per Cl. No. 9 of IEC 60079-0:2011.
- External surface temperature rise test.
- Hydraulic pressure test is done after the 'explosion' test.

6.14.1.4 Non-sparking (Ex 'n A')

- Creepage and clearance as per IEC 60079-15:2010.
- Control on motor bearing design (specially sleeve bearing design).
- Stator ignition tests (steady state only).
- Maintain minimum allowed radial air gap as per Cl. No. 8.7 of IEC 60079-15:2010.
- Maintain creepage and clearance of the neutral point connection as per Cl. No. 8.6 of IEC 60079-15:2010.
- Assessment for possible air gap sparking as per Cl. No. 8.8.3 of IEC 6079–15:2010 (however, for S1 or S2 duty motors this is not required).
- Non-metallic material on which the type of protection depends selection, specification, and thermal endurance test as per IEC 60079-1:2011, Cl. No. 7.
- Surface temperature rise test of the motor (both internal and external) during the motor normal service condition.
- Use of IECEx approved space heaters.
- Use of IECEx approved terminal blocks for winding RTD and bearing RTD terminations.
- Selection of fasteners and holes for fasteners as per Cl. No. 9 of IEC 60079-0:2011.
- Selection of metals for the metallic enclosures and metallic parts of enclosures as per Cl. No. 8 of IEC 60079-0:2011.
- Use of IECEx approved cable glands.
- Any other accessories required to be used in the motor should be IECEx approved.
- Ventilation (ventilation openings, requirements of cooling fan) shall be as per Cl. No. 17.1 of IEC 60079-0:2011.
- Torque test for bushing as per Cl. No. 26.6 of IEC 60079-0:2011.
- Use of anti-loosening, anti-vibration-type connection for all electrical connections, as per Cl. No. 14 and 15 of IEC 60079-0:2011.

6.14.1.5 Increased Safety (Ex 'e')

All the above requirements for Ex 'n A' are applicable, with the following additions:

- Stator ignition test (Impulse ignition + steady-state ignition test).
- Locked rotor surface temperature rise test for the stator as well as rotor.
- Creepage and clearance as per IEC 60079-7:2006.
- Equipment with rated voltage more than 11 kV not under the scope of standard or certification.

- Additional electrical connection requirement as per Cl. No. 4.2 of IEC 60079-7:2006.
- Assessment for possible air gap sparking as per Cl. No. 5.2.4.3 and potential stator winding discharge risk assessment is required to be done and special measures like pre-start purging are to be employed. Provision for installation in motor manufacturer scope. Pre-start purging equipment and its employment are in user scope. Manufacturer should duly inform all the relevant inform related to purging to the user in the form of 'Special Condition of Safe Use' for all the hazardous area motors required to be operated with VFD, and the specific VFD+Motor combination is required to be tested for the temperature rise test, as per IEC 6079-0:2011 at the normal service conditions. This test is required to be duly witnessed by the ExCB and then issues a supplementary certificate to the original IECEx certification for the product.

6.14.2 Factory Approval

- Quality Management System (QMS)
  The quality management system in the factory for the hazardous area motors should be in accordance with IEC 80079-34:2011, over and above ISO 9001:2008. The QMS is required to be duly audited by the ExCB or the authorized person nominated by the ExCB. Based on successful completion of the audit, the ExCB issues the Quality Assurance Notification/Quality Assurance Report (QAN/AAR).
- Test Plan approval (OD024) agreement)
  The test plant is also audited by the ExCB as per requirements of IECEx directive OD024, and upon successful completion of the audit, the ExCB makes OD024 agreement with the manufacturer.

  Based on the above two documents, the ExCB thus approves the manufacturing facility, i.e., issues factory approval.

The regional requirements and procedure for certification all over the world are more or less same with the procedure as described above for IECEx certification. The difference is that, it is governed and monitored by the different regional government authorities or, the authorities duly approved by the respective governments. Hence, preparing for this IECEx certification makes an electrical equipment manufacturer prepared at least with respect to the design, manufacturing, and quality requirements for the most of the other regional certifications also. However, repeating the same tests again and again is a costly and time-consuming affair. Moreover, getting the design and test results approved by the different authorities makes the process more complex and makes business economically inviable. Hence, it would be wise for a motor manufacturer to get itself prepared for the most stringent as well as common requirements in design, quality, and production process and then venture for the different certifications.

Appendix I

FORM NO.: (CIMFR: DQM: FLP02: FR-01)

APPLICATION FOR PROTOTYPE/REVALIDATION CERTIFICATION OF FLAMEPROOF (Ex d) ELECTRICAL APPARATUS/ ENCLOSURE and INGRESS PROTECTED APPARATUS

Application Ref. No.: Date:

1. Applicant/Authorized representative
   Name and Full Address
   Telephone
   Fax
   e-mail
2. Manufacturer, if different from above
   Name and Full Address
   Telephone
   Fax
   e-mail
3. Name / Title of apparatus
4. Details of electrical ratings:
5. Degree of Ingress Protection (if applicable):
6. Type/ Cat No/ Model No/ Sl. No. of the Apparatus:
7. Surface temperature class: T6 T5 T4 T3 T2 T1
7. Gas Group (please tick [√]): Gr. I Gr. IIA Gr. IIB Gr. IIB+H2 Gr. IIC
8. (a) Material of construction

   (b) Whether casting or of fabricated type

   (c) In case of light alloy please provide test report of chemical composition:

   (d) Frictional incendivity test required Yes No
9. Approximate Weight of the empty enclosure with cover (in Kg):

10. Details of motor:

Type of motor	Frame Size	HP/KW	Volt	Current	No. of poles	Speed	No. of phase	Insulation class

11. (a) Details of Glass (for enclosures containing glass):

Sr. No.	Size	Shape	Type	Thickness	Cemented Path

(b)

Name of sealing compound	Make	Color	Operating temperature min. and max. (0C)

12. Details of lighting fittings (Type and rating of the lamps/tube):

13. Details of Flameproof Apparatus:

Sl No.	Name of Enclosure	Dimen-sion	Volume* (in liters)		Nature of Joint	No. and size of bolts	Max No. of Aperture on Cover	Max No. of terminals	Max No of Entry and Type on each side of enclosure			
			Gross	Net					SIDE			
									A	B	C	D

14. Details of Cable gland/Nipple/Stopping plug/Sealing conduit:

Sl. No.	O.D.	I.D.	Detail of gasket	Type	Nature of thread	Axial length of thread	Material of construction

15. List of Drawings Submitted with the application:

Sl. No. Title Drawing No. Sheet No Revision

No Date

16. Name and number of standard with year to which the apparatus is claimed to comply:

17. State any other relevant information:

18. State whether your company is registered under S.S.I. unit: Yes/No.
(If yes please state S.S.I. Certificate No. and also
submit a Xerox copy of S.S.I. unit certificate.)

19. (a) State whether test fee has been deposited: Yes/No.
If 'Yes' please quote demand draft No. with date, amount and name of bank.

(b) State whether sample(s) equipment have been submitted:
If 'Yes' Please give the particulars: Yes/No.
In case 'No' please indicate likely date of submission.

20. Whether a duplicate certificate is required and, if so, No. of additional certificate copies:

* Please note that the test charges shall be calculated on the basis of declaration of the necessary gross volume of the enclosure of the equipment and total number of enclosures involved. In absence of the above information, test charges shall be calculated only on receipt of the equipment and in that case test charges calculated by CIMFR shall be considered as final and binding on the company. No negotiation will be entertained afterwards in this respect.

Note-1: In case of light alloy a spare cover of the enclosure should be submitted for frictional incendivity test if required.

Note-2: See guidelines of testing attached with this form.

DECLARATION

I/We have read the information circular, understood and hereby agree to be bound by the requirements asked for.

I/We agree that the sample/equipment submitted for test at CIMFR shall be collected back by me/us within 90 days on receipt of Test Report failing which CIMFR would no way be responsible for safe custody of the sample/equipment and CIMFR has sole right to dispose off the sample/equipment by public auction without any further notice to me/us.

Signature of the applicant: Office Seal
(Authorized Signatory*)

Name:

Designation:
Contact no.:

Date:

Appendix II

FORM NO.: (CIMFR: DQM: FLP02: FR-02)

APPLICATION FOR SUPPLEMENTARY / CHECK / BATCH TESTING

(Please strike out whichever is not applicable)

Application Ref. No.: Date:

1. Applicant
   Name and Full Address
   Telephone
   Fax
   e-mail
   CIMFR Registration No.:
2. Manufacturer, if different from above
   Name and Full Address
   Telephone
   Fax
   e-mail

3. Title of apparatus, type identification/Designation with details of electrical rating:
4. Cat No/ Model No/ Sl. No.:
5. (a) Gas Group (please tick [√]): Gr. I î Gr. IIA î Gr. IIB î Gr. IIC î
   (b) Zone (where it is proposed to be installed): Zone 0 î Zone 1 î Zone 2 î
6. Standard reference:
7. Temperature Class (please tick [√]): T1 î T2 î T3 î T4 î T5 î T6 î
8. Material of Construction
   (in case of light aluminum alloys, specify the composition of alloy):

9. Proto type Test Certificate No. and Date issued earlier
(Including Amendment/Addendum, if given):
(Attach Xerox copies)

10. Supplementary Certificate Numbers and Dates, if issued earlier:

11. State type of certificate required: Supplementary ǐ Check ǐ Batch Testing ǐ
(Please tick [√])

12. In case of Supplementary/Check test, state
(a) Details of the variations:
(b) List of Drawing Submitted with the application:
Sl. No. Title Drawing No. Sheet No Revision No
Date

Note: In case of Gr. IIC applications no supplementary certificate is issued.

13. Details of Glass window of the modified product (if applicable):

Nos.	Size	Shape	Type	Thickness	Cemented Path	Sealing Material

14. Details of modified apparatus (if applicable):

Sl No.	Name of Enclosure	Dimension	Volume*		Type of Cover/Joint	No. and size of bolts	Max No. of Aperture on Cover	Max No. of terminals	Max No of Entry and Type on each side of enclosure SIDE			
			Gross	Net					A	B	C	D

15. Details of modified Cable gland/Nipple/Stopping plug/Sealing conduit (if applicable):

Sl. No.	O.D.	I.D.	Detail of gasket	Type	Nature of thread	Axial length of thread	Material of construction

16. Details of modified Flame arrestor/flash back arrestor (if applicable):

Sl. No.	Type	Directional / Bi-directional	Flanged diameter/ size	Mesh size of sintered	Gap between parallel plates	Material of arrestor element	Material of arrestor body

17. In case of Batch Testing, state (100% OR As per amendment):
    (a) Lot Size and Serial Numbers:
    (b) Serial Numbers of Batch units submitted for testing:

18. State any other relevant information:

19. State whether your company is registered under S.S.I. unit: **Yes/No.**
    If yes please state S.S.I. Certificate No. and also submit a Xerox copy of S.S.I. unit certificate.

20. (a) State whether test fee has been deposited: **Yes/No.**
    If 'Yes' please quote demand draft No. with date, amount and name of bank.
    (b) State whether sample(s) equipment have been submitted:
    If 'Yes' Please give the particulars : **Yes/No.**
    In case 'No' please indicate likely date of submission.

21. Whether a duplicate certificate is required and, if so, No. of additional certificate copy

* Please note that the test charges shall be calculated on the basis of declaration of the necessary gross volume of the enclosure of the equipment and total number of enclosures involved. In absence of the above information, test charges shall be calculated only on receipt of the equipment and in that case test charges calculated by CIMFR shall be considered as final and binding on the company. No negotiation will be entertained afterwards in this respect.

Note-1: Please see the **Annexure-I** for permissible variations.

Note-2: Please strike out whichever is not applicable.

DECLARATION

I/We have read the information circular, understood and hereby agree to be bound by the requirements asked for.

I/We agree that the sample/equipment submitted for test at CIMFR shall be collected back by me/us within 90 days on receipt of Test Report failing which CIMFR would no way be responsible for safe custody of the sample/equipment and CIMFR has sole right to dispose off the sample/equipment by public auction without any further notice to me/us.

Signature of the applicant: **Office Seal**

Name:

Designation:

Date:

Appendix III

FORM NO.: (CIMFR: DQM: FLP02: FR-04)

TITLE:APPLICATIONFOR INTRINSIC SAFETY/EXPLODER /SHOT FIRING CABLE/HEATING CABLE TEST.

Application Ref. No.

Date:

1. Name of apparatus and type/cat. number if any
2. Indigenously manufactured or imported type.
3. The operating instruction of the equipment, a brief write up regarding principle of operation and description of equipment must also be furnished.
4. Name and Address of Applicant.
5. Name and address of manufacturer (if different from item No. 4)
6. Zone and Gas Group
7. Temperature Class
8. List of Drawing of apparatus and circuit diagram with all component values in detail enclosed with applications for examination.

Sl. No.	Drawing No.	Date	Rev. No.	Title

9. Standard specifications to which sample to be tested.
10. Whether a duplicate certificate is required and, if so, No. of additional certificate copy
11. Whether enclosure of the equipment is FLP/DP/DT, specify with material of construction(in case of light aluminum alloys, specify the composition of alloy. Frictional incendivity test is necessary)
12. For exploder, specify the No. and Type of detonator to be fired using the exploder.

(i) Discharge Time

(ii) Residual Charge

(ii) DC output Voltage

(iv) Type of Battery and rating

(v) Permitted type/non permitted type

13. For Intrinsic Safety of Circuit/Apparatus specify.

(i) No. of Intrinsically safe circuits and No. of non-intrinsically safe circuits which are associated with the former.

(ii) Operating Voltage of the circuit.

(iii) Safety fuses rating and Type.

(iv) Open voltage of power supply

(v) Short circuit current

(vi) Total inductance of the circuit

(vii) Total capacitance of the circuit

14. For inductive components used in the circuit:

(i) No. of turns, gauge of wire and material of all windings.

(ii) Details of insulation

(iii) D-C resistances of the windings.

(iv) Dimensions and materials of all cores in the magnetic circuits.

15. In case of Shot firing cable, give the complete specification like single shot or multi shot type

16. In case of Heating cable, give the complete specification, whether

(i) constant watt series (ii) constant watt parallel (iii) self limiting,

also mention voltage and output rating and the maintenance temperature and withstanding temp.

17. State any other relevant information

18. State whether your company is registered under S.S.I. unit :

Yes/No. If yes please state S.S.I. Certificate No. and also

submit a xerox copy of S.S.I. unit certificate

19. (a) State whether test fee has been deposited : Yes/No.

If 'Yes' please quote demand draft No. with date, amount

and name of bank.

b) State whether sample(s) equipment have been submitted:

If 'Yes' Please give the particulars : Yes/No.

In case 'No' please indicate likely date of submission.

Note: Please strike out whichever is not applicable.

D E C L A R A T I O N

I/We have read the information circular, understood and hereby agree to be bound by the requirements asked for.

I/We agree that the sample/equipment submitted for test at CMRI shall be collected back by me/us within 90 days on receipt of Test Report failing which CMRI would no way be responsible for safe custody of the sample/equipment and CMRI has sole right to dispose off the sample/equipment by public auction without any further notice to me/us.

Signature of the applicant: Office Seal

Name:

Designation:

Date:

Guidelines for Prototype Certification as per IS/IEC 60079-11: 2011

1. Manufacturer should ensure before applying for prototype testing as per IS/IEC 60079-11: 2006 for intrinsic safety testing, that the design of intrinsic safety with applicable safety factor for particular gas group and Level of Protection is assessed for their product.

Drawings of the Intrinsic Safety product/equipment should confirm the compliance with IS/IEC 60079-11: 2006 and other relevant standards. The following information must be mentioned in the drawings:

Enclosure/housing dimensions detail of the equipment

Complete detail of PCB Layout, Track, Material, Thickness, CTI etc.

Complete Block diagram, if it is an Associated apparatus or more than one product contains in the testing Sample.

Complete detail of the Transformer, Relay, Fuse, Inductive Coil, Cell/batteries etc. as per requirement of IS/IEC 60079-11: 2011.

Details of Creepage distances, clearances etc.

The Entity parameters of the apparatus, if any.

Total capacitance, inductance, L/R ratio, etc as per requirement of IS/IEC 60079-11: 2006.

markings and warning markings and its position and method of marking as given in the relevant standards.

size of the drawings should be minimum A3size. Drawings should also be very clearly and visible.

Appendix IV

(FORM NO.: (CIMFR: DQM: FLP02: FR-03)

9.3 TITLE: APPLICATION FOR PVC/STEEL CORD BELTING/ HOSES / BRATTICE SHEET / VENTILATION DUCTING TEST.

Application Ref. No. Date:

1. Name of Applicant with Full address:
   Phone No., Fax No. E mail
2. Name of Manufacturer:
3. Belt Description:
   (a) Type Designation:
   (b) Identification Color:
   (c) Width of belt:
   (d) Thickness of Belt :
   (e) Thickness of bottom cover:
   (f) Thickness of top cover:
   (g) Material of
      (i) Cover:
      (ii) Fabric:
      (iii) Impregnating fire resistant compound:
   (h) No of steel cord per width in meter:
   (a) No. of Piles: (in case of PVC belting)
   (b) Tensile strength in KN/cm
4. For Hose:
   Type of Hose with material specification:
   Internal Diameter:
   Outer Diameter:
5. For Brattice Sheet:
   (a) Type of Sheet with material specification:
   (b) Sheet Thickness:
   (c) In case of impregnated sheet (i) material of impregnation (ii) material of fabric
   (d) In case of unsupported plastic, material of plastic:

6. Standard specifications to which sample to be tested:

7. Tick on the Test to be carried out:

   (a) Drum Friction Test

   (b) Electrical Resistance Test (Antistatic Test)

   (c) Propane Gallery Test

   (d) Flammability Test (Sprit Burner Test)

8. Whether a duplicate certificate is required and, if so, No. of additional certificate copy

9. Any other relevant information, if any:

   Note: Please strike out whichever is not applicable.

Signature of the Applicant

Name and Designation: **Office Seal:**

Date:

Appendix V

Guidelines for preparation of drawings:

1. **General:**

Following points should be take care of during preparation of flameproof drawing.

(i) Flameproof drawing of electrical equipment should clearly show the flamepath, gap, clearance, creepage, wall thickness as per IEC, code requirement.

(ii) Preparation of drawing may be any convenient scale and it should clearly represent the design and construction of the whole equipment.

(iii) Flameproof drawing should be signed by the manufacture/applicant of that product.

(iv) The electrical component of flameproof products which do not directly affect the flameproofness, may be shown in outline.

(v) All dimensions such as flamepath, gaps, the diameter and clearance of shaft and spindles bot holes size and spacing and wall thickness should be given tolerances if any upon the nominal dimensions.

(vi) In case of range of flameproof motors, which are identical in general design but differ in dimensions then one representative drawing will suffice, provided for each size in the range are set out in a schedule upon the drawing.

(vii) If different parts of flameproof enclosure have already been certified such as plug and socket coupler or an externally mounted instrument etc should be given on the general arrangement drawing.

(viii) Different types of cable attachment should be shown in the general arrangement drawing.

(ix) Each flameproof drawing should be clearly numbered and dated and there should be providion for insertion of revision or issue numbers and dates to cover modifications in design.

(x) Drawing showing general arrangement of product should bear the following particulars:

a) A declaration should be made by the product manufacturer that this particular flameproof product complies with the IEC 6079-1:2014 as soon as flameproofness is concerned.

b) Proper catalogue number, model number or any variations should be declared by the manufacturer.

c) Details of rating of the flameproof products.

2. Sectional views

a) Sectional views of flameproof enclosure should be given by the manufacturer it should include flamepath and gaps at each joints, the dimensions of bushings and shaft gland.

b) The dimensions and position of all the holes for the insertion of screws, bolts, studs or rivets in flanges or elsewhere of flameproof enclosure should be shown and their dimensions should be figured.

c) If joint of flameproof enclosure is spigot then length of flange portion and spigot portion should be clearly mentioned.

d) The insertion of insulated conductor through a hole in terminal box and through bushing in main flameproof enclosure should be shown in detail and fully dimensioned.

e) If any venting or drawing device inserted in the wall of flameproof enclosure dimension should be shown clearly.

3. Plan views

It is required to show variation in breadth of opposed joint surfaces where dimensions varies and to show the position of securing studs, bolts or screws if their fittings are not properly.

4. Supplementary drawings

a) If some components of the flameproof products are not clearly visible in the flameproof drawing so additional drawing should be prepared on a larger scale system.

b) If interlocking system are present in the design of the product then it should be clearly mentioned on the primary drawings and shown detail in the supplementary drawings.

5. Material of construction

a) The material of construction of flameproof products, screws, bolts and studs used for attaching parts or components and bushing to insulating material or cement that is applied to close the openings should be clearly mentioned in the flameproof drawings.

b) If welding is used for the attachment of any parts to the main flameproof enclosure, this should be shown clearly on the drawings.

Appendix VI

FORMAT NO.: (CIMFR: DQM: FES02: F-01:REV-01)

Format for certification of Flameproof product

Equipment ID No.: ULR No.: Code No.

Test and Assessment Report No.:	*Dated:*
Application Ref. No.:	*Dated:*

1. Applicant :
2. Manufacturer :
3. Equipment :
4. Type of protection :
5. Electrical ratings :
6. Temperature Class :
7. Degree of Ingress Protection :
8. **Material of Construction** :
9. Description of the equipment :

A.

Enclosure Type	Volume (cc).		Min. Wall Thickness (mm)	Nos. and Size of Bolts/Fasteners in (mm)
	Gross	Net		

B.

Glass details	Max no. of aperture on cover

10. Nature of Flameproof Joint: Type of joints and gaps (Spigot joint, Threaded joint, Cylindrical joint for Gas Gr. IIB)

Sr. No	Location of flamepath	Type of Joint	Min. length of flamepath (in mm)		Max. gap (in mm)/no. of threads (pitch)	
			Req.	Spec.	Req.	Spec.

Max No. and size of external/exit cable entries:

11. Name plat and warning inscription

12. Drawings:

Sl. No.	Drg. No.	Title	Rev.	Sheet	Date

13. Any other relevant information:

14. Declaration by the Applicant/Manufacturer:

15. Documents/Samples Submitted :

(i) Application form

(ii) Drawings

(iii)Prototype sample

16. Condition of sample received: Good

17. Compliance of prototype or sample with documents: The test sample of electrical equipment submitted for the type tests complies with the manufacturers documents referred above.

Note: CIMFR has however not checked and tested the compliance of the equipment to any standard other than the above standards.

SCOPE OF THE TEST CERTIFICATE

The Test Certificate issued by CIMFR certifies that the equipment has been found to comply with the definition of Flameproof-Weatherproof equipment contained in the relevant

Standard specifications. They do not vouch for the quality of the equipment in any other respect. The present test report pertains to the sample received only.

This Institute reserves the right to review, amend or withdraw this Test Report at any time if considered necessary in the interest of safety.

REPORT OF TEST

Date of Test:

Test Equipment Used:

Result #A: Tests as per IS/IEC 60079-0: 2017:

Type Tests			
Clause	Tests	**Remarks**	Results (Complies, P- Pass, NA-Not Applicable, F-Fail) References
1	Scope		
2	Normative references		
3	Terms and definitions		
4	Equipment grouping		
5	Temperature		
5.1.1	Ambient Temperatures		
6	Requirements for all electrical equipment		
6.6	Electromagnetic and ultrasonic energy radiating equipment.		
7	Non-metallic enclosures and non-metallic parts of enclosures		
7.1.2.4	Material used for cementing / Sealing		
7.2	Thermal Endurance		
8	Metallic Enclosure and metallic part of Enclosure		
9	Fasteners		

10	Interlocking devices		
11	Bushings		
12	Reserved for future use		
13	Ex components		
14	Connection facilities		
15	Connection facilities for earthing and bonding conductors		
16	Entries into enclosures		
17	Supplementary requirements for rotating machines		
18	Supplementary requirements for switchgear		
19	Reserved for future use		
20	Supplementary requirements for plugs and socket outlets and connectors for field wiring connection		
21	Supplementary requirements for luminaries		
22	Supplementary requirements for cap lights and hand lights		
23	Equipment incorporating cells and batteries		
24	Documentation		
25	Compliance of prototype or sample with documents		
26	Type tests		
26.3	Tests in explosive test mixtures		
26.3	Tests for Flameproof (Ex ‘d’) protection		
26.4	Test of enclosures		
26.4.2	Resistance to Impact		
26.4.3/ 26.4.4	Drop test		

26.4.5	Degree of protection (IP) by Enclosures		
26.4.5	Test of IP of equipment		
26.5	**Thermal tests**		
26.5.1	Temperature measurement		
26.5.2	Thermal shock test		
26.6	Torque Test for bushings		
26.6	Torque test for bushings		
26.7	Non- metallic enclosures or non-metallic parts of enclosures		
26.8	Thermal endurance to heat		
26.9	Thermal endurance to cold		
26.10	Resistance to UV light		
26.11	Resistance to chemical agents for Group I Equipment	-------------	
26.12	Earth continuity test in non-metallic enclosure		
26.13	Surface resistance test of parts of enclosures of Non-metallic materials	-------------	
26.14	Measurement of capacitance		
26.15	Verification of ratings of ventilating fans		
26.16	Alternative qualification of elastomeric sealing O-rings		
27	**Routine tests**		
28	Manufacturer's responsibility		
29	Marking		
30	Instructions		

Result #B: Tests in explosive mixtures: Tests for type of protection Ex 'd' as per IS/IEC 60079-1:2014.

Type Tests	
Clause No.	Description and relevant tests
1	Scope
2	Normative references
3	Terms and definitions
4	Level of protection (Equipment Protection Level, EPL)
5	Flameproof joints
6	Sealed joints
6.1	Cemented joints
6.1.2	Mechanical strength
7	Operating rods
8	Supplementary requirements for shafts and bearings
9	Light transmitting parts
10	Breathing and draining devices which form part of a flameproof enclosure
11	Fasteners and openings
12	Materials
13	Entries for flameproof enclosures
14	Verification and tests
15	Type Tests
15.1	General
15.2	**Tests of ability of the enclosure to withstand pressure**
15.2.1	**General**
15.2.2	Determination of Explosion Pressure (Reference Pressure)

15.2.3	Overpressure Test	
15.3	Test for non-transmission of an internal ignition	
15.3.1	General	
15.3.2	Electrical equipment of Groups I, IIAand IIB	
15.3.3	Electrical equipment of Group IIC	
15.4	Tests of flameproof enclosures with breathing and draining devices	
15.4.1	General	
16	Routine tests	
17	Switchgear for Group I	
18	Lamp holders and lamp caps	
19	Non-metallic enclosures and non-metallic parts of enclosures	
19.1	General	
19.2	Resistance to tracking and creepage distances on internal surfaces of the enclosure walls	
19.3	Requirements for type test	
19.4	Test of erosion by flame	
20	Marking	
20.2	Caution and warning markings	
21	Instruction	

Result # B1: Determination of Explosion pressure (Reference Pressure Test) *:

<table>
<tr><td colspan="7">Test Condition</td></tr>
<tr><td colspan="3">Type of Test</td><td>Gas Group</td><td colspan="2">Gas Mixture % in Air</td><td>No. of Tests</td></tr>
<tr><td colspan="3">Preliminary Test (reference pressure)</td><td>I</td><td colspan="2">10% Methane in air</td><td>Three</td></tr>
<tr><td colspan="2">Test Ref. No.</td><td>Ignition</td><td>Sensor Position</td><td>Sensor Position</td><td>Max. Pressure millisecond</td><td>Remark</td></tr>
<tr><td>Body</td><td>Body</td><td>PPM/</td><td></td><td>PPM/</td><td></td><td>Satisfactory</td></tr>
</table>

**Results shown for the pressure time curve (enclosed) are the highest recorded value obtained in the test.*

Result #B2: Overpressure Test (Static Method): Overpressure test is conducted as per Cl. of IS/IEC 60079-1: 2014.

Enclosure	Test reference no.	Over pressure (kg/cm^2) Maintained	Remark

Result #B3: Test for Non transmission of internal ignition:

Test Condition					
Enclosure	Test Ref. Nos.	Gas Group	Gas Mixture % in Air	No. of Test	Result

CONCLUSION:

Reported By ( )

Checked and Approved By ( )

Appendix VIA

CIMFR	सीएसआईआर – केंद्रीय खनन एवं ईंधन अनुसंधान संसथान CSIR- CENTRAL INSTITUTE OF MINING AND FUEL RESEARCH (वैज्ञानिक तथा औधोगिक अनुसंधान परिषद्) (COUNCIL OF SCIENTIFIC AND INDUSTRIAL RESEARCH)	NABL

TYPE EXAMINATION CERTIFICATE

FORMAT NO.: (CIMFR: DQM: FLP02: F-01: REV-01)

FLAMEPROOF AND EQUIPMENT SAFETY

Certificate No.: CMF 22 INEx 0067 Dated: 10/02/2022

ULR No.: TC596222000000079 (Issue No. 0)

1. Applicant : M/s. XYZ Private Limited
2. Manufacturer : Same as Above
3. Equipment : Flameproof and Weatherproof ....
4. Designated By : Cat. No.: ....
5. Type of Protection and EPL: Ex db Gas Group ..... EPL ....

Fig. 1 Equipment Certificate issued by CIMFR, Dhanbad

Appendix VII

DGMS APPLICATION FORM FOR APPROVAL OF SAFETY EQUIPMENT FOR USE IN MINES

(This application alongwith the drawings and other enclosures should be submitted to the Director-General of Mines Safety, Dhanbad – 826001 (Bihar), in duplicate)

PART – I

1. Particulars of Applicant:
2. Full Postal address:
3. Telegraphic address:
4. Address of location of factory:

GENERAL INFORMATION TO BE FURNISHED:

1. Date of establishment of business in India:
2. Nature of the concern – whether Public Ltd. Company, Private Ltd. Company, Partnership or Hindu Undivided Family Concern:
3. Name of Directors, Partners, Proprietor or Karta as the case may be:
4. Capital Investment –
   i) Machinery & equipment (Details of machinery to be attached):
   ii) Land & Buildings or rent premises:
5. Registration number allotted by the State & Director of Industries:
6. BIS Certification Licence No.:

PART – II

1. Name of the equipment of material:
2. Description of the equipment or material:
3. Production Capacity:
4. Actual Production, if any:
5. Estimated production in ensuing year:
6. Price of the product:
7. Delivery time from the date of order:

PART – III

1. Specifications of the equipment or material) Name the I.S. or B.S. Specification(s):
2. Drawings of the equipment or material:
3. Reference to Indian standard or any other standard to which the equipment or material conforms:
4. Broad specifications of the equipment or material:
5. Test Report on the equipment or material:
6. Particulars of raw materials and components used in manufacture:
7. Instructions for operation:
8. Instructions for maintenance:
9. One complete sample of the equipment or material, if possible.

Place: Signature:

Date:

Stamp:

References

1. Ramachandiran R (2014) Rules and regulations of electrical equipment used in underground coal mines and in hazardous areas of oil mines. In: 1st international seminar on design, development, testing and certification of ex-equipment, DTEX – 2014. Dhanbad
2. Srivastava PC (2014) Approval of electrical equipment used in hazardous areas – role of PESO. In: 1st international seminar on design, development, testing and certification of ex-equipment, DTEX – 2014. Dhanbad
3. Vishwakarma RK, Singh AK, Ahirwal B, Sinha A (2010) Explosion pressure development and temperature rise classification of low rating flameproof electric motors. Int J Petroleum Sci Technol 4:1–7. ISSN 0973-6328
4. IEC 60079-1:2014 – Electrical apparatus for explosive gas atmospheres Part-1 flameproof enclosure 'd'
5. Coal Mine Regulations 1957
6. Indian Electricity Rules 1956
7. Sarkar SK (1988) Flameproof equipment design, construction, certification use and maintenance 1988

Chapter 7
Initiation and Prevention of Explosion Through Non-electrical Means

7.1 Introduction

We know that when coal is extracted from underground coalmines, methane is liberated, and that methane combining with air creates an explosive atmosphere. A similar situation occurs during the drilling of oil when gases such as hydrogen, acetylene, ethylene, and propane are emitted, mixing with the air to create an explosive atmosphere. In such an explosive atmosphere, we cannot use normal electrical equipment but can use a special type of equipment, and this equipment is called Ex equipment. Whenever electrical equipment comes in contact with flammable media, an explosion may occur. This may result in the death of a large number of people in underground coal mines and other related industries. It is not only the failure or fault of electrical equipment used in areas containing flammable gases, vapors, and liquids that can present a fire and explosion hazard, but there may also be some instances of fire and explosion caused by non-electric sources such as flame arresters, light aluminum alloy materials, and non-sparking tools. This chapter discusses in detail the design, working principles, types, applications, testing procedures, and uses of flame arresters. Moreover, the mechanism of testing and certification of non-sparking tools and the non-incendivity properties of light metal alloys (LM6) are also discussed in this study.

7.2 Flame Arrester

7.2.1 Flame Arrester and Its Operation

A flame arrester is a safety device intended to allow gas, vapor, and liquid to flow through it, but to prevent the transmission of flames through it. In general, these are fitted to the openings of enclosures or pipework. It has been widely used for a long time in the petrochemical, fertilizer, and pharmaceutical industries. The operation of

© The Author(s), under exclusive license to Springer Nature Singapore Pte Ltd. 2023
A. Kumar Singh, *Explosion-Proof Equipment in Hazardous Area*,
https://doi.org/10.1007/978-981-99-2516-2_7

a flame arrester is very straightforward. Known for its ability to stop the transmission of flames and to withstand the explosion pressure generated by shock waves, it is able to stop the spreading of flames. A famous chemist and Professor at the Royal Institution in the U.K., Sir Humphry Davy, discovered the operation principle of flame arresters in 1815 [1]. As a passive device, it has no moving parts during operation. However, if an explosive mixture is ignited and the flame attempts to return to the gas source, it will prevent the flame from traveling.

Working Principles:

It is a passive device (without moving parts), which consists of passageways of apertures that allow gases, vapors, and liquids to pass through it, but not flames. As the flame front travels through the passageways of the apertures, it absorbs heat from the flame front traveling at subsonic speeds, that is, it acts as a heat sink, thus lowering the air mixture concentration below its auto-ignition temperature and preventing the flame from surviving [2].

The size of the flame arrester channel is determined by the flammability of the fuel mixture. In underground coal mines, coal dust clouds and firedamp are highly explosive, so the mesh of a Davy lamp must be very closely spaced. Ideally, the wire mesh of the flame arrester should be protected from damage caused by being struck by another object. Wires should be shifted carefully so as not to create an opening that will allow the flame to spread beyond the barrier (Fig. 7.1).

Basic Design of Flame Arrester:

Flame arrester is having three parts:

- Body
- Arrester element
- End flanges.

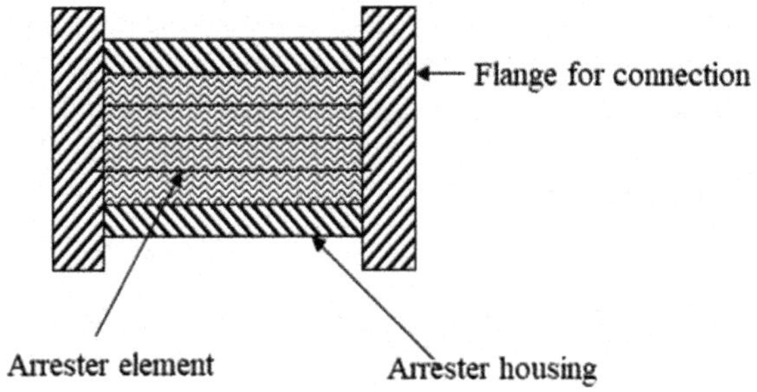

Fig. 7.1 General layout of a flame arrester

7.2.2 *Different Types of Flame Arrester*

An explosive mixture present in an explosive atmosphere can burn in a variety of ways. Consequently, flame arresters can be divided into different types. The combustion process is affected by mechanisms such as pre-compression, pressure waves, and flame propagation speed. The following are some combustion processes for flame arresters:

- Explosions are oxidation processes in which temperature, pressure, or both abruptly increase simultaneously [3, 4].
- In deflagration, explosion pressure propagates at a subsonic speed (273 m/s) [3].

When choosing flame arresters, it is important to distinguish between the following [4]:

1. Deflagration in the atmosphere (Fig. 7.2): This type of explosion occurs in the open air without a noticeable increase in explosion pressure.
2. Pre-volume deflagration (Fig. 7.3): This type of explosion occurs inside a vessel and is caused by an internal ignition source.

(A) **In-line deflagration** (Fig. 7.5)**:** An explosion is developed within the pipe that moves with flame propagating speed along the axis.
(B) **Stabilized burning** (Fig. 7.4)**:** It is short time burning close to the flame arrester, and endurance burning is for an unlimited time.

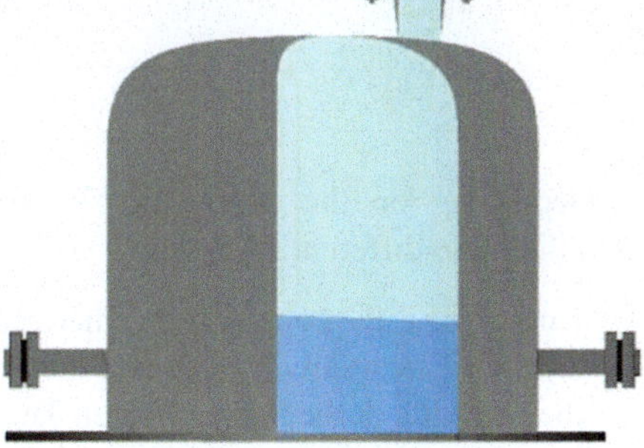

Fig. 7.2 Atmospheric deflagration (A)

Fig. 7.3 Pre-volume deflagration (B)

Fig. 7.4 Stabilized burning (D)

Depending on the combustion process and installation, flame arresters are classified into different types.

1. **End-of-line flame arrester**: There is only one pipe connection on this flame arrester. Essentially, it prevents the transmission of flame from the open air to the vent side of the flame arrester. The design of flame arrester is made such that it prevents the flame and maintains the mechanical integrity when traveling the flame front. It operates on atmospheric pressure and is generally installed on top of storage tanks, vessels, etc. It is recommended that a distance of 1.5 m (5 feet) be maintained between the flame arrester and the ignition source (Fig. 7.6).

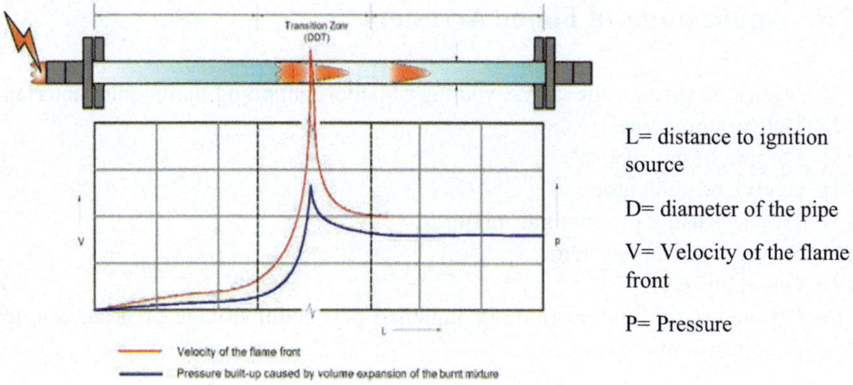

Fig. 7.5 In-line deflagration unstable/stable

Fig. 7.6 End-of-line flame arrester

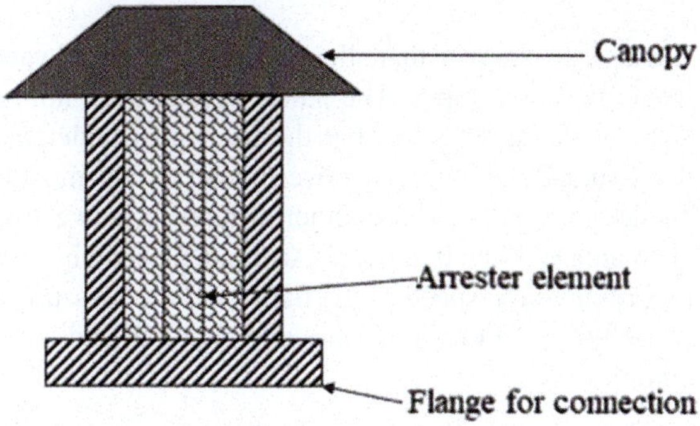

2. **In-line-deflagration flame arrester**: This flame arrester is also known as a deflagration and detonation flame arrester. The device is installed between the pipes in order to prevent flames from passing through. The purpose of this type of flame arrester is to collect gas from liquids and solids. If high speed (above supersonic) flame passes through the in-line-deflagration, then its design is not perfect to arrest the flame and maintain its structural integrity.
3. **In-line-detonation flame arrester**: An in-line-detonation flame arrester is installed in a piping system, regardless of the distance from the source of ignition and the configuration of the same size piping system. It is prepared and tested to stop traveling flame fronts in low, medium, and high deflagration ranges. The device must be installed in piping with a diameter equal to or less than that of the piping used in the test.

In accordance with the test agency's guidelines, the arrester can operate at elevated operating pressures and with different types of gases.

7.3 Applications of Flame Arresters

(1) Open vent pipes, vent valves, venting of tanks containing flammable materials.
(2) Delivery pipe lines.
(3) Purging of gas mains.
(4) Acetylene generators.
(5) Furnaces using pre-mix gas mixtures.
(6) Solvent recovery systems.
(7) Gas analyzers.
(8) Petroleum, oil, and gasoline or liquefied petroleum storage or processing or piping systems.

7.4 Selection of a Flame Arrester

It is a passive device that allows liquid, gases, and vapors to flow but prevents flames from igniting in pipes. The flame arrester need not be tested or certified to be suitable for the application, but the user must understand the actual area of use. For the assurance of safety, the user/manufacturer must be aware of flame propagation phenomena. There is a distinct difference between an explosion that occurs in an open area and one that occurs within a closed pipe system. Flame arresters are used according to the speed of the flame and the pressure waves in the flame.

Before selecting a flame arrester, the end user must follow the following guidelines.

- The user must ensure that the flame arrester is used in specific areas for which it has been tested and certified by the testing authority.
- It is important to note that the capability of a flame arrester is dependent on the operating pressure of the vapor, liquid, or gas in the pipe system.
- Connections between piping systems and flame arresters must be flameproof.
- The length of pipe and its diameter must not exceed the dimensions for which the flame arrester was tested and certified.
- It is possible that bends, elbows, and valves in piping may compromise the flame arrester's performance.
- The flame arrester's operating temperature must be limited as certified by the certifying agency.
- It is essential that the end user/manufacturer ensures that the flow rate of the gas or vapor is clearly specified.
- The material of construction of the flame arrester and its orientation in the pipeline must be clearly specified.

7.5 Installation of Flame Arrester

- There are no moving parts in flame arresters. The device is passive and does not require calibration or modification.
- It is necessary to remove the flange protection and packing material from the arrester.
- A flame arrester should always be mounted vertically or horizontally.
- The bolts on the flange should be tightened, and the gaskets should be compatible with the service conditions.

7.6 Maintenance of a Flame Arrester

Maintenance of flame arresters should include the following considerations:

- The maintenance of flame arresters should be carried out according to the prescribed maintenance schedule.
- Entire flame arrester must be taken out during removal process of connection, stud/bolts from pipeline.
- In the case of welded flame arresters, the arrester element cannot be removed.
- Testing authorities/manufacturers should visually inspect the arrester element for damage to the winding. It should be replaced immediately if it appears to be damaged.
- It is recommended that gaskets be inspected periodically and replaced if necessary.
- A light source should be viewed through the passages of flame arresters to determine if there is a blockage of the opening.
- A solvent wash may be used to clean the arrester element, followed by compressed air flow and purging with compressed air.

It can be hazardous if the end user/manufacturer fails to select the flame arrester as per the requirements. The end user/manufacturer should be aware of the characteristics of flame transmission. This includes the effects of temperature and pressure of gas or vapor as well as the size of the piping system. When using flame arresters, it is important to adhere to the limitations stated in the International Electro-technical Commission (IEC) standard. Table 7.1 illustrates the mixture of air gases used in the testing of the detonation and deflagration types of flame arresters.

7.7 In-Line Flame Arrester (Deflagration Test)

Figure 7.7 shows the testing procedure of in-line flame arrester. A spark plug works as an ignition source fitted in the center of the blind flange. The size of flame arrester connector and pipe diameter should be same for hydrogen–air mixture (IIA, IIA1, IIB3, IIB2, IIB1), unprotected length of pipe should not be less than 10xD and should

Table 7.1 Specification of gas–air mixtures for detonation and deflagration tests

Range of application (marking)		Test mixture			
Gas group	Mixture of MSEG	Type of gas	Purity of gas (volume %)	Gas in atmosphere (volume %)	Safe gap of air gas mixture (mm)
IIC	< 0.50	Hydrogen	≥ 99	28.5 ± 2.0	0.31 ± 0.02
IIB3[b]	20.65	Ethylene	298	6.6 ± 0.3	0.67 ± 0.02
IIB2[b]	20.75			5.7 ± 0.2	0.73 ± 0.02
IIB1[b]	20.85			5.2 ± 0.2	0.83 ± 0.02
IIB[b]	20.50	Hydrogen	299	45.0 ± 0.5	0.48 ± 0.02
IIA[b]	> 0.90	Propane	295	4.2 ± 0.2	0.94 ± 0.02
IIA1	21.14	Methane	298	8.4 ± 0.2	1.6 ± 0.02

not be greater than 50xD, and for (IIC and IIB) hydrogen mixture, pipe diameter should not greater than 30xD. For hydrocarbon air mixture (IIA, IIA1, IIB3, IIB2, IIB1), protected pipe length LP shall be 50xD, and hydrogen–air mixture (IIC and IIB) protected pipe length will be 30xD.

As illustrated in Fig. 7.7, there are two flame arresters installed on the unprotected side of the pipe in order to measure the flame velocity.

Distance between two flame detectors is 'b' pressure recording system (limiting frequency 2 100 kHz) which is used to record the pressure developed during explosion fitted in pipe toward unprotected side at a distance 'a' in the Fig. 7.7.

1. Ignition source toward blind flange.
2. Inlet for mixture.
3. Lu and 'D' are length and diameter of unprotected pipe.
4. Flame detector to measure the flame.
5. In-line deflagration flame arrester.
6. Peizo electric pressure transducer to measure the pressure.
7. Flame detector.
8. Diameter 'D' and length LP are the diameter and length of protected pipe, respectively.

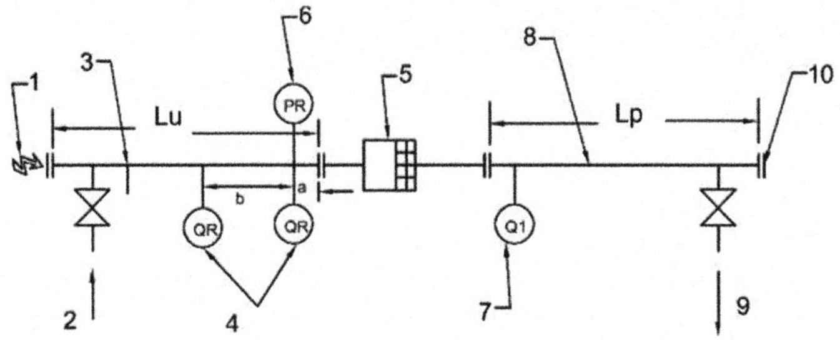

Fig. 7.7 Arrangement for in-line flame arrester for deflagration test

9. Outlet for mixture.
10. Blind flange.

$$2D \geq a(\pm 10\% \text{ max.} \pm 50\,\text{mm}), \text{ but } 9550\,\text{mm}$$
$$3D \geq b$$

As per Table 7.1, a test mixture should be prepared, and six consecutive tests should be carried out, without flame transmission. Transmission of flame detected by flame detector [5] on the protected side.

7.8 End-of-Line Flame Arrester

As shown in Fig. 7.8, the arrangement for testing the end-of-line flame arrester for deflagration tests should be measured from the top of the flame arresters.

A gas mixture shall be prepared as per Table 7.1 and placed in a polythene bag and ignited by a spark plug during the testing. Two tests are performed at each ignition point, resulting in a total of six tests. In order to detect flames on the protected side, a flame detector is used.

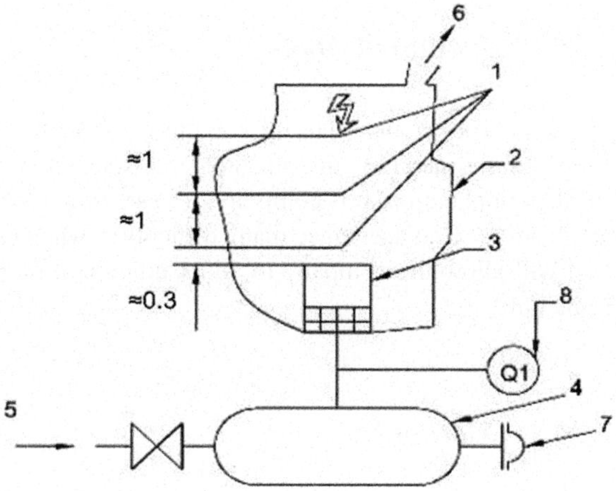

Fig. 7.8 Arrangement for end-of–line flame arrester for deflagration test: key. (1) Source of ignition, (2) Polythene bag, (3) End-of-line flame arrester, (4) Explosion pressure—resistant vessel, (5) Mixture inlet, (6) Mixture outlet, (7) Bursting diagram, and (8) Flame detector for measuring flame

7.9 Non-sparking Tools

7.9.1 Introduction

A non-sparking tool does not emit a high-temperature spark. The sparks emitted by non-sparking tools are cold and have very low heat contents, which cannot ignite the explosive atmosphere. In general, steel tools made from ferrous alloys and Chrome Vanadium can emit sparks capable of igniting flammable materials. Among the materials used in making non-sparking tools are Aluminum Bronze and Beryllium copper. CSIR-Central Institute of Mining and Fuel Research, Dhanbad tests and certifies both alloys according to IS 4595 [6]. The Beryllium copper alloy is the strongest and has the most desirable mechanical properties for making non-sparking tools [7]. In addition to Beryllium copper, Aluminum Bronze is another copper alloy used for non-sparking tools. It is more economical than Beryllium copper, and its use is widespread in end user applications. There are some companies that offer tools made of brass that are too soft to be used for wrenches. When purchasing such tools, end users should be vigilant. Any industry relies heavily on hand tools. In several industries, such as petrochemicals, fertilizers, steel, chemicals, and pharmaceuticals, non-sparking alloys should be used to prevent explosions.

7.10 Selection of Tools

Different types of materials are used to manufacture hand tools for different applications. It is crucial for any stakeholder to select non-sparking tools that are suitable for their application. It is common for users to purchase the incorrect tools for their applications. End users face many difficulties when using these tools. Proper selection will allow the company to adopt non-sparking tools and ensure safety at the same time.

7.10.1 Material

Any non-sparking material can be used to manufacture non-sparking safety tools. Below is a description of the chemical composition of one of these non-sparking materials [5]:

Constituent	Percent
Iron	3.5–5.5
Nickel	4.5–6.5
Manganese	1.5, Max

(continued)

(continued)

Constituent	Percent
Aluminum	8.5–10.5
Copper	Remainder

Impurities in the non-sparking material, whose composition is shown in Sect. 7.10.1, shall not exceed the following limits:

Constituent	Percent
Silicon	0.25
Magnesium	0.05
Lead	0.05
Tin	0.10

7.11 Non-sparking Tools Test

Non-sparking materials are tested in accordance with IS: 4595:1969. CSIR-CIMFR, Dhanbad, maintains a very specific test setup for the material sample. The environment inside the test chamber consists of 20 cm^3 gasoline and air mixture containing of 50% oxygen (Ambient temp. of 22 °C). A collision is created between the sample and the fixed pole having rotating disk (10,000 rpm) of the apparatus. During the collision, there should be no spark which can explode the explosive mixture available in the test setup (Fig. 7.9).

7.11.1 Use of Non-sparking Tools

Following the selection of the correct tools for their workplace, it is equally important to be aware of the limitations of each tool. Standard hand tools are designed for specific purposes, but if used for any other purpose, they can cause problems. As an example: Double ended spanners should only be used manually and should not be used for hammering or pipe work requiring more torque, as this may result in the misuse of the tool. As well, hammering a screw driver or hammering a wrench is another misuse of tools.

Thus, it is clear that this tool is not appropriate for your application, and it is being misused in order to accomplish the task. The life of tools is reduced when they are used in such a manner. A usage guideline should be provided by the manufacturer to customers in order to ensure that the tool is used properly.

Fig. 7.9 Test setup for testing non-sparking tools

There are many accidents that may occur if customers do not use the proper tools in the industry. There are a large number of fires and accidents in the industries caused by frictional sparks caused by the improper use of non-sparking tools.

Operators and maintenance personnel should follow the following three steps in order to mitigate the risk of fire.

1. Identify the extent of the hazard correctly. In other words, the flammable zone.
2. Choose tools do not spark and are suitable for hazardous areas.
3. Operators and maintenance personnel are required to apply tools correctly.

Taking these three steps will certainly improve your plant's safety.

7.12 Frictional Incendivity Test

7.12.1 Introduction

A flameproof enclosure is designed in such a way that if a fire occurs within the enclosure, the flame or combustible product should not communicate with the outer flammable media, and its mechanical integrity should remain intact during the explosion. Steel, cast iron, welded sheets, and steel construction are the most common materials used for flameproof products. The light metal alloy is also used as material of construction of flameproof product if its composition is such that it does not give rise to incendive sparking. However, the material chosen for the construction of flameproof enclosures should be resistant to corrosion by agents such as water. It is common practice in foreign countries to construct flameproof enclosures using FRP or GRP. However, in India, electrical equipment constructed using FRP or GRP is not accepted by statutory bodies.

Because of their attractive characteristics, light metal alloys such as LM6, LM4, and LM0 are widely used in various fields, such as lightness and strength. In underground coal mines, electrical equipment made of light metal alloy has been subject to serious explosions when it strikes rusty steel plates or other hard metals [8].

In underground coalmines, the hazard arises when the spark energy is produced during strikes with metal to metal in flammable gas. The spark energy is capable of igniting the flammable media. Hand drills used by miners in underground coal mines may cause explosions if they slip and fall on rusty steel plates. We can produce a dangerous spark energy when we hammer on an aluminum painted rusty steel plate structure [9–11]. Under different conditions, spark energy production varies widely.

Experimentally, it was observed that the spark energy required to ignite methane–air mixtures in underground coal mines is of the order of 10 mJ.

7.12.2 Light Metal

Generally, aluminum, magnesium, titanium, and cerium are called light metal alloys, and all have the capability of giving incendive sparks. The magnesium alloy is more dangerous than the aluminum alloy, as its susceptibility to ignition increases with magnesium content. Under the updated code IEC 60079-0:2017 [12], magnesium percentage should not exceed 7.5 when LM6 light metal alloy is used as a construction material.

7.12.3 Laboratory Studies on Light Metal Alloy

The ignition possibilities of different light metal alloys have been investigated extensively in laboratory [5, 7]. As a result of the experiment, it was observed that the probability of ignition is 19% if the magnesium concentration in the alloy is 11%, but only 2% if the magnesium concentration is 0.05% [13].

7.12.4 Different Coating Material

There have been many attempts to produce light metal alloys free of hazards [14] by adding various additives to the metals.

1. As an anti-frictional material in light metal alloys, polyterafluoroethylene and molybdenum disulfide are used.
2. Tin and lead having lower melting point.

Table 7.2 Influence of magnesium content on probability of ignition [13]

Alloy	Mg	Si	Cu	Ignition probability percent
LM-0M	0.05	0.5	0.2	2
LM-6M	0.05	10.19	0.11	2
LM-4M	0.15	4.0–6.0	2.0–4.0	4
LM-16	0.4–0.6	4.5–5.5	1.0–1.5	3
WP	1.59	0.50	4.02	2
LM-14	5.02	0.08	–	9
WP	10.73	0.05	–	19
LM-5M LM-10 W				

3. Anti-oxidants, which act as inhibitors of the thermite type of reaction.

There is some coating material, such as araldite, that reduces the likelihood of incendiary sparks occurring. It was observed in a study that the percentage of ignition was reduced from 84 to 34% by using araldite as a coating material. Other coaling materials, such as polythene, epoxy resin, molded rubber, neoprene, solder, shaft sprays of zinc, solder, stove enamel, or lead, have been used.

A study found that protection varies from coating to coating and that such protection lasts only as long as the coating is intact.

Table 7.2 and Fig. 7.1 show magnesium's influence on LM6 ignition probability.

7.12.5 Present Status Regarding Use of (LM-6) Light Metal Alloys

Many countries permit the use of light metal (LM6) with some restrictions on the content of magnesium, titanium, and aluminum. The Bureau of Indian Standards (BIS) has now been harmonized with the International Electro-technical Commission (IEC). Under these codes, alloys used in the construction of electrical equipment for underground coal mines, i.e., Group I, shall not contain by weight [8] (Fig. 7.10).

1. A total of more than 15% aluminum, magnesium, and titanium in the alloy.
2. A total of more than 6% of magnesium and titanium in the alloy.

In accordance with IEC 60079-0:2017, magnesium should not exceed 7.5% by weight of the alloy used in the construction of electrical equipment for Gr II, i.e., surface industries.

In surface industries, such as gas groups IIA, IIB and IIC, alloys like LM6, LM5, and LM4 may be used in the construction of electrical equipment. The code does not provide much information regarding the composition of the alloy. Prior to the harmonization of the Bureau of Indian Standards, LM0, LM6, and LM4 alloys used for the construction of electrical equipment in gas group IIA, IIB, and IIC should

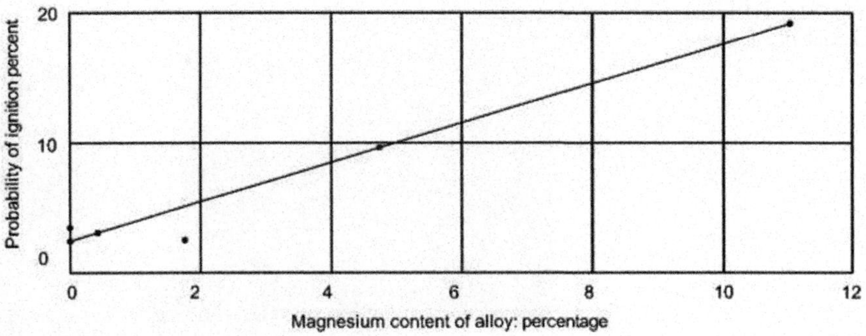

Fig. 7.10 Variation of probability of ignition with magnesium content

have a magnesium percentage of less than 0.2%. However, after the harmonization of the BIS code as per IEC, a magnesium percentage of 7.5 is allowed in alloys such as LM0, LM6, and LM4.

7.12.6 Frictional Incendivity Test

There are three main methods that are adopted for testing the light metal alloy as follows [15].

1. The drop test in which one material strikes another either in free fall or by hammer blows.
2. Frictional smear test in which softer light metal leaves a smear on rusty steel plate and it is struck on sliding or grazing blow.
3. Rubbing friction test in which moving parts of a power driver machines are striked by a stationary object.

During a frictional incendivity test, a piece of electrical equipment made of light metal alloy (LM6) is first cut, then screwed onto a 16 kg cylindrical brass weight and dropped from a vertical height of 4 m with 16 kg of additional weight on a rusty inclined steel plate placed at an angle of 45 °C to the vertical. Between 35 and 55° from vertical, the angle of impact is not critical. The impact will be minimum, and the friction will be maximum in this situation. When a sample of light metal is dropped on a rusty steel plate, the combustion of the eroded particles helps ignite the flammable gas mixture. This results in a flash. An increase in magnesium content in light metal alloys results in a decrease in the incendivity of the flash produced by impact. The lower explosive limit of methane is 5%. And the upper explosive limit is 15%. The most explosive limit is 8.5%. The mixture of methane and air is most easily combustible when it contains 6.5% methane. In accordance with the code requirements, the test is repeated five times. The alloy is acceptable if, in none of the tests, an ignition of 21% hydrogen–air mixture occurs (Fig. 7.11).

Fig. 7.11 Frictional incendivity test setup

It is difficult to use LM6 as a material for the construction of flame-resistant products during mass production. Many prototype samples have passed the frictional incendivity test, but the alloy used in mass production has failed to meet the quality requirements. Therefore, manufacturers and users should exercise caution. When manufacturers purchase light metal alloys in bulk, they should ensure that the composition and incendivity of the metal is tested. Additionally, while end users are purchasing flameproof products made of light metal, they should always select one or two samples at random and send them to the testing laboratory for testing. It is a simple and inexpensive test. Moreover, a high-quality protective coating is also necessary.

References

1. Scheed A, Chakravarty N (2014) History of development of FLP/IS electricals and emerging trend in development. In: 2nd international seminar, DTEX-2014
2. Singh AK, Flame arrester in hazardous area. HRD, course for ONGC exclusive
3. ISO 16852:2008 Flame arresters- performance requirements, test methods and limits for use
4. Davies M, Kurzel A, Heidermann T (2014) Flame arresters. In: 2nd international seminar on "design, development, testing and certification of ex-equipment. DTEX 2014 at Kolkata, India

5. Introduction to non-sparking tools posted on 12th April, 2021
6. NFPA 69, standard on explosion protection
7. IS: 4595-1969 general requirements for non-sparking tools
8. Hawkins Colliery, Kinnock Chase 1950 and Harden Colliery 1953
9. Fire – (1938) Standard for portable fire extinguishers, vol 31, pp 156
10. Thomas TSE (1941) Iron coal trades review, vol 143, pp 210
11. Kingman FET, Coleman EH, Roginsky ZW (1952) J Appl Chem 2:449
12. IEC 60079-0:2017 Electrical apparatus for explosive gas atmosphere Part- 0: General requirements
13. Sarkar SK (1988) Flameproof equipment-design. Construction, certification and maintenance April
14. Latin A (1959) Colliery Eng 36:397–404 and 440–446
15. Bailey-Trans Inst Min Eng 118:223 (1958–1959)

Chapter 8
Selection and Installation of Electrical Equipment in Hazardous Areas

8.1 Introduction

In selecting electrical equipment for hazardous areas, one must have an understanding of combustible media (gas grouping), the extent of hazards (zone), temperature classification (T-Class), and environmental conditions. Additionally, it is critical to observe proper safety requirements like potential equalization, electrical protection, emergency switch off, etc., when installing the equipment. In order to facilitate the selection of Ex equipment, the International Electro-technical Commission (IEC) has introduced the concept of zones and equipment protection levels (EPLs). EPLs are required on all electrical equipment under this concept. Since Ex equipment with multiple protection will also carry an EPL, there will be no confusion regarding its suitability for use in a particular zone (Fig. 8.1).

During the selection of Ex equipment for hazardous areas, it is taken into account that the temperature class of Ex equipment should not exceed the auto-ignition temperature (AIT) of gases and vapors present in the hazardous area.

As per the requirements, Ex equipment is tested for gas groups I, IIA/IIB, and IIC combustible media. Gas group IIC poses the greatest risk, and gas group I poses the least risk. Equipment tested for gas group IIC can also be used in gas groups IIA/IIB and I (underground coal mines), but equipment tested for gas group I cannot be used in gas groups IIA/IIB or IIC. It is necessary to select electrical equipment that is dust and liquid-resistant.

Electrical equipment should be used within its rated power, current, frequency, duty rating, and any other characteristics. It should also be ensured that the temperature classification of the electrical apparatus has been established for the above ratings.

It is the sole responsibility of the product manufacturer to manufacture and obtain certificates for Ex equipment. Once purchased, the safety of these Ex equipment is covered by the end user. End users are responsible for selecting, installing, inspecting, and commissioning Ex equipment. There is therefore a distinct role for both manufacturers and users of Ex equipment.

© The Author(s), under exclusive license to Springer Nature Singapore Pte Ltd. 2023
A. Kumar Singh, *Explosion-Proof Equipment in Hazardous Area*,
https://doi.org/10.1007/978-981-99-2516-2_8

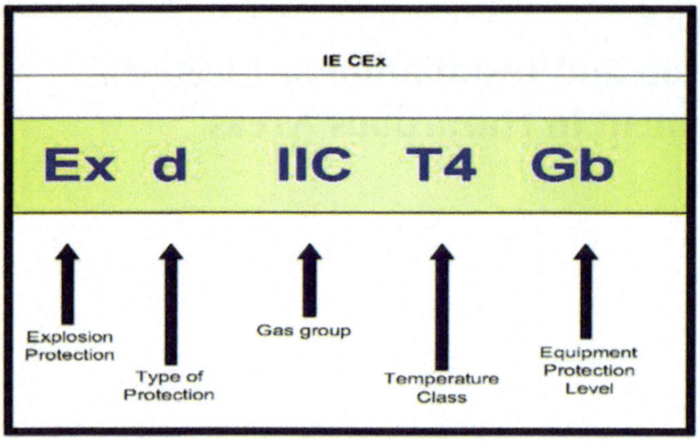

SR EN ISO 80079-36		Directive 2014/34/UE (GD no. 245/2016)		SR EN 60079-10-1 SR EN 60079-10-2
EPL	Group	Equipment Group	Equipment Category	Zones
Ma	I	I	M1	Not Applicable
Mb			M2	
Ga	III	II	1G	0
Gb			2G	1
Gc			3G	2
Da	III		1D	20
Db			2D	21
Dc			3D	22

Fig. 8.1 Marking and relation between equipment protection levels (EPLs) and zones

Bureau of Indian Standards (BIS), Directorate General of Mine Safety (DGMS), and Petroleum and Explosives Safety Organization (PESO) are the technical bodies in India that provide advice on the adoption of technical standards and the appropriate use of Ex equipment in hazardous areas such as underground coal mines and petroleum industries such as petrochemicals, refineries, and fertilizers.

The purpose of this chapter is to provide information regarding the selection of Ex equipment based on various criteria such as zone, T-class, gas group, and environmental conditions and also provides the information about installation.

8.2 Electrical Apparatus Selection (Excluding Cables and Conduits)

A proper selection of Ex equipment requires the following information [1]:

- Hazardous Area Classification
- Temperature class and ambient temperature
- Gas group
- Environmental conditions.

8.2.1 Selection of Ex Equipment for Hazardous Areas Based on Explosion Probability

The probability of explosion (P_{ex}) can be defined based on the probability of occurrence of the explosive atmosphere (P_{a}) and the probability of formation of an ignition source (P_{s}) [2],

$$P_{ex} = P_a \times P_s$$

It is clear that areas having less probability of occurrence of an explosive atmosphere (P_{a}) will have minimum chances of explosion. The simultaneous occurrence of an explosive atmosphere and ignition source is also important. From Table 8.1, it is clear that the protection techniques Ex 'n' and Ex 'e' are basically designed to prevent any ignition source from arising. Similarly, flameproof protection is designed to prevent any ignition from spreading. In the case of pressurized protection, the equipment is designed to prevent the flammable mixture from reaching a means of ignition.

The use of electrical equipment in a particular explosive atmosphere depends upon the type of atmosphere in which it is tested. Table 8.2 shows the different Ex equipment with examples and uses in surface industries.

The protection is achieved by suitably constructing and designing the electrical equipment. In the background of each protection, the main thing is to prevent the contact among explosive mixture, ignition sources, and supportive environment, generally oxygen as shown in Table 8.3.

Table 8.4 shows the relevant codes, protection concepts, equipment protection levels, and types of protection appropriate for use in each zone.

Table 8.1 Ex equipment and use in hazardous area as per explosion probability

Type of protection	P_a	P_s	P_{ex}	Use of type of protection in hazardous areas					
				Zone 0	Zone 1	Zone 2	Degree I	Degree II	Degree III
Ex 'd'	≤ 1	1		x	√	√	√	√	√
Ex 'i'	≤ 1	0	0	a√	√	√	√	√	√
Ex 'e'	<< 1	0	0	x		√	√		
Ex 'p'	0	1	0	x	√	√			
Ex 'o'	0	1	0	x	√	√			
Ex 'm'	0	1	0	x	√	√			
Ex 'q'	0	1	0	x	√	√			
Ex 'n'	<< 1	0	0	x		√			

Table 8.2 Different Ex equipment corresponding to zone

Type of protection	Examples of Ex equipment	Condition/area of use	Zone
Increased safety non-sparking	Terminal boxes/motors/ lighting fittings	Where explosive atmosphere is not present or presents for very short period	Zone 2
Intrinsic safety	Very low-power apparatus Sensors/telecommunication system/RFIDS	Where explosive atmosphere is present continuously	Zone 0 and also in zone 1 and zone 2
Encapsulation	Miniature devices/small components	Where explosive atmosphere is normally present	Zone 1 and zone 2
Pressurization	Control panels/motors/ transformers	Where explosive atmosphere is normally present	Zone 1 and zone 2
Oil immersion	Transformers	Where explosive atmosphere is normally present	Zone 1 and zone 2
Sand filled	Transformers	Where explosive atmosphere is normally present	Zone 1 and zone 2
Flameproof	Junction boxes/control panels/ switching devices/gate end boxes/motors/lighting fittings	Where explosive atmosphere is normally present	Zone 1 and zone 2

Table 8.3 Type of protection for electrical equipment and method of protection

Method	Type of protection
By eliminating arc, spark and hot surface explosion is prevented	Ex 'e' increased safety Ex 'n' non-sparking
By restricting the output, energy explosion is prevented	Ex 'i' intrinsic safety
By keeping the flammable media out of reach of ignition source	Ex 'm' encapsulation Ex 'p' pressurization ex 'o' oil immersion ex 'q' powder/sand filled
Designed to prevent any ignition from spreading	Ex 'd' flameproof enclosure

8.2.2 *Equipment Protection Level (EPL)*

Equipment protection levels (EPLs) are based on a grading system. In gassy mines, the level of protection is determined by its potential to ignite, and it creates a difference between explosive gas, flammable dust, and explosive atmosphere.

In order to facilitate the selection of Ex equipment, IEC has introduced the concept of EPL. According to this concept, all Ex equipment must be marked with an EPL [4]. This concept is very useful as Ex equipment with multiple protection will also carry EPL so there is no confusion about its suitability for use in a particular zone. Table 8.5 shows selection of electrical equipment as per EPL.

8.2.2.1 Equipment for Use in Zone 0

Equipment for use in zone 0 should have equipment protection level (EPL) 'Ga'.

This equipment has a very high level of protection, ensuring that it does not act as a source of ignition in the event of a malfunction.

It is suitable for installation in flammable media of Group II. Under IEC 60079-11:2011 [5], intrinsically safe equipment of category 'ia' falls into this category.

8.2.2.2 Equipment for Use in Zone 1

If electrical equipment is designed and constructed for use in zone 0 or any electrical equipment having EPL 'Gb' can be used in zone 1 hazardous areas.

As well as being suitable for use in Group II explosive atmospheres, this equipment provides a very high level of protection. Electrical equipment of this type does not have a source of ignition during normal operation or during expected malfunctions.

(a) As per IEC 60079-1, flameproof equipment Ex 'd' comes under this category [3].
(b) As per IEC 60079-2, purged protected equipment Ex 'p' comes under this category [6].

Table 8.4 Selection of electrical equipment according to zone/EPL [3]

Type		Relevant code		Protection concept	EPL and suitable for use in zone	
Gas	Dust	Gas	Dust		Gas	Dust
		IEC 60079-0		General requirements		
Ex d		IEC 60079-1		Flameproof	Gb (1, 2)	
Ex px	Ex pD	IEC 60079-2	IEC61241-4	Purge protected	Gb (1, 2)	Db (21, 22), Dc (22)
Ex py					Gb (1, 2)	
Ex pz					Gc (2)	
Ex q		IEC 60079-5		Powder filled	Gb (1,2)	
Ex o		IEC 60079-6		Oil immersed	Gb (1,2)	
Ex e		IEC 60079-7		Increased safety	Gb (1, 2)*	
Ex ia	IEC 60079-11	Intrinsically safe	Ga (0, 1, 2)	Da (20,21,22)		
Ex ib			Gb (1, 2)	Db (21, 22)		
Ex ic			Gc (2)	Dc (22)		
Ex nA				Non-sparking	Gc (2)	
Ex nC				Enclosed break	Gc (2)	
Ex nR		IEC 60079-15		Restricted breathing	Gc (2)	
Ex ma		IEC 60079-18		Encapsulation	Ga (0, 1, 2)	Da (20,21,22)
Ex mb					Gb (1, 2)	Db (21, 22)
Ex mc					Gc (2)	Dc (22)
	Ex ta		IEC 60079-31	Dust ignition protected		Da (20,21,22)
	Ex tb					Db (21, 22)
	Ex tc					Dc (22)

Table 8.5 Selection of electrical equipment in hazardous areas as per EPL

Zone	0	1	2	20	21	22
EPL	Ga	Ga, Gb	Ga, Gb, Gc	Da	Da, Db	Da, Db, Dc

(c) As per IEC 60079-5, powder filling Ex 'q' comes under this category [7].
(d) As per IEC 60079-6, oil immersion Ex 'o' comes under this category [8].
(e) As per IEC 60079-11, intrinsic safety Ex 'i' comes under this category [5].
(f) As per IEC 60079-18, encapsulation 'm' comes under this category [9], and
(g) Electrical heat tracers and equipment certified for zone 1 hazardous area comes under this category.

8.2.2.3 Equipment for Use in Zone 2

The electrical equipment having equipment protection level 'Gc' is used in zone 2 in addition to electrical equipment that can be used in zone 0 and zone 1.

The equipment is suitable for use in explosive atmospheres of gas group II, having an enhanced level of protection. The equipment does not serve as an ignition source in normal operation and may have some additional protection to ensure that in the event of regular occurrences of flammable media, it does not function as an ignition source.

(a) As per IEC 60079-15, non-sparking Ex 'n' comes under this category [10].
(b) As per IEC 60079-11, intrinsic safety Ex 'i' comes under this category [5].
(c) As per IEC 60079-7, increased safety Ex 'e' comes under this category [11].
(d) Electrical heat tracers and equipment, which are certified for use in zone 2 area.

8.2.3 Selection Based on Temperature Classification

During area classification, we specify the temperature class of gasses and vapors present in the hazardous area based on their auto-ignition temperature as per IEC 60079-20-1 [12] (Table 8.6).

During the selection of Ex equipment, it must be ensured that the temperature class of the equipment does not exceed the auto-ignition temperature (AIT) of gases and vapors present in the hazardous area.

Table 8.6 Temperature classification of gasses and vapors based on their auto-ignition temperatures

Temperature class	Auto-ignition temperature of gases and vapors (°C)
T1	> 450
T2	> 300
T3	> 200
T4	> 135
T5	> 100
T6	> 85

For example, hazardous areas containing hydrocarbons will have a temperature class of T3 as determined by the area classification. Thus, Ex equipment of temperature classes T3 and lower, i.e., T4, T5, and T6, can be used. This is because their maximum surface temperature will not reach the ignition temperature of any gases or vapors present in such areas.

By default, all Ex equipment is suitable for ambient temperatures ranging from − 20 to + 40 °C. In such a case, the marking of the electrical apparatus does not specify the ambient temperature range. If the apparatus is suitable for lower/higher ambient temperatures, the same is indicated on the Ex equipment.

As a result of the above, all Ex equipment shall be used only within the ambient temperature range of − 20 to + 40 °C unless otherwise indicated. Nevertheless, if the manufacturer of the equipment requires a lower/higher ambient temperature, it may be specified by the user to ensure that the manufacturer provides appropriate Ex equipment.

8.2.4 Selection as Per Gas Group

IEC 60079-0:2017 specifies that

- Electrical apparatus of type of protection Ex d, Ex i, Ex nC, and Ex nL shall be marked and used for suitable gas group sub-division IIA, IIB, or IIC.
- Electrical apparatus of type of protection Ex e, Ex m, Ex p, Ex q, Ex o, Ex nR, and Ex nA shall be suitable for Group II gasses as a whole and accordingly will not be marked as IIA, IIB, or IIC.

IEC 60079-0:2017 specifies that all type of protection shall be marked for gas group sub-division IIA, IIB, or IIC.

Electrical apparatus suitable for gas group IIC can also be used with gas groups IIB and IIA under less hazardous conditions. Furthermore, electrical apparatus that is suitable for gas group IIB may also be used for gas group IIA under less hazardous conditions.

Listed in Table 8.7 are the categories of combustible media according to the NEC and IEC, as well as representative gasses within each category.

Hydrocarbon and other industries generally contain H2 that belongs to gas group IIC along with gasses belonging to gas groups IIA and IIB. Among the other gases

Table 8.7 Sub-division of combustible media as per NEC and IEC and its respective representative gas

Representative gas	Sub-division as per NEC	Sub-division as per IEC
Acetylene	A	IIC
Hydrogen	B	IIC
Ethylene	C	IIB
Propane	D	IIA

in gas group IIC are carbon di-sulfide and acetylene. In these industries, acetylene is generally absent, while carbon disulfide is found only in Rayon factories. An acetylene environment does not permit flange-type construction beyond 500 cc. For panels and power junction boxes, flange-type enclosures are much easier to install and maintain than spigot-type enclosures. In industries that utilize hydrogen and other gases belonging to gas groups IIA and IIB, there is a concept of using Ex d IIB + H2 flange-type control panels and power junction boxes. For this reason, during the process of area classification, if no acetylene or carbon di sulfide is present but only hydrogen is present, the area will be categorized as IIB + H2 so that flange-type enclosures can be used.

8.2.5 Environmental Condition

Electrical apparatus should be selected based on their ability to withstand environmental conditions, including chemical presence, vibration, process heat, humidity, and power quality.

8.3 Electrical Apparatus Installation (Excluding Cables and Conduits)

In order to install electrical products in hazardous areas, the end user or manufacturer should be aware of the existing worldwide codes applicable to hazardous areas. Table 8.8 compares the Indian Standards with those of other countries.

8.3.1 General Requirements

Electrical installations in hazardous areas are required to comply with the following requirements [13].

- Electrical equipment should be used within the limits of its power, current, frequency, duty rating, as well as other characteristics. For each of the above ratings, it should also be ensured that a temperature classification has been established for the apparatus.
- In accordance with IEC 60079-17:2007 [14], an initial inspection should be conducted following the completion of the erection.
- It is recommended that electrical apparatus be located in areas that are non-hazardous or least hazardous.
- It is only permitted to transport or change fluorescent tubes if the hazardous area has been declared to be free of IIC gases.

Table 8.8 Comparison of Indian Standards with those of other countries

Subject	IEC	Australia, New Zealand	Europe	USA		Canada	
				FM	UL	Ex zone model	Ex class/division model
Ex: general requirement	60079-0:2017	AS/NZS 60079:200 8	EN 50014	FM 3600	–	CSA 79-0-95	–
Flameproof ex 'd'	60079-1:2014	AS/NZS 60079.1 AS 2380.2	EN 50018	FM 3615	UL 2279, Pt.1 UL 1203	CSA—E79-1	CSA C22.2NO.30
Intrinsic safety ex 'I'	60079-11:2006	AS/NZS 60079.11 AS 2380.7	EN 50020	FM 3610	UL 2279, Pt. 11 UL 913	CSA—E79-7	–
Increased safety ex 'e'	60079-7:2015	AS/NZS 60079.7 AS 2380.6	EN 50019	–	UL 2279, Pt. 7	CSA—E79-7	
Pressurized enclosure ex 'p'	60079-2:2014	AS/NZS 60079.7 AS 2380.4	EN 50016	FM 3620	NFPA 496	CSA—E79-2	
Non-sparking ex 'n'	60079-15:2015	AS/NZS 60079.15 AS 2380.9	EN 60079-15	FM 3611	UL 2279, Pt. 15	CSAE79-15	
Encapsulation ex 'm'	60,079-18:2014	AS/NZS 60079.18	EN 50028	–	UL 2279, Pt. 18	CSA—E79-18	
Oil filled ex 'o'	60079-6:2000	AS/NZS 60079.6 AS 2380.7	EN 50015	–	UL 2279, Pt. 6	CSA—E79-6	
Powder filling ex 'q'	60079-5:2015	AS/NZS 60079.5	EN 50017	–	UL 2279, Pt. 5	CSA—E79-5	

- Low-pressure sodium vapor lamps should not be used in hazardous areas due to the risk of ignition from free sodium in a broken lamp.

8.3.2 Documentation in Order to Ensure Proper Installation of Ex Equipment, the Following Documents May Be Required

- Documents related to area classification.
- Instructions for installation and connection.
- Conditions specific to Ex equipment bearing a U and X certification.
- Document describing a system that is intrinsically safe.
- Information pertaining to inspections.
- Calculations, such as the purging rate for instruments or analyzer houses.
- Information required for repairs under IEC 60079-19:2006 [15].
- Certificate of authenticity for Ex equipment.

8.3.3 Installation of Intrinsically Safe Products

When installing intrinsically safe apparatus, the specific characteristics of the gas or vapor involved are taken into account in relation to the ignition current or minimum ignition energy. Table 8.9 describes the gas group, representative gasses, and corresponding ignition energies.

8.4 Safety Requirements

8.4.1 Potential Equalization

Installations in hazardous areas require potential equalization. For TN, TT, and IT systems, all exposed and extraneous conductive parts must be connected to the equipotential bonding system. A bonding system may consist of protective conductors, metal conduits, metal cable sheathing, steel wire armoring, and metal parts

Table 8.9 Gas group, representative gases, and their ignition energies [16]

Explosion Group	I	IIA	IIB	IIC
Gas	Methane	Propane	Ethylene	Hydrogen
Ignition energy (mJ)	0.28	0.26	0.095	0.018

of structures, but not neutral conductors. The connection must be secure against self-loosening.

- Conductive parts are exposed to the equipotential bonding system and are in metallic contact with structural components or piping do not require separate connections.
- Metallic enclosures of intrinsically safe apparatus are not required to be connected to the equipotential bonding system unless required by the apparatus documentation or to prevent static buildup.
- Cathodic protection installations should not be connected to an equipotential bonding system unless the system is specifically designed to do so.

8.4.2 Electrical Protection

These requirements apply to all circuits with the exception of intrinsically safe circuits.

The wiring and apparatus should be protected from overloads, short circuits, and earth faults. A short circuit and earth fault protection device should not automatically close in the event of a fault.

Whenever one or more phases are lost, protection against overload and overheating must be provided. A few examples of overload protection devices for rotating electrical machines are as follows:

- A current dependent time lag protective device monitoring all three phases set at rated current of the machine, which will operate in 2 h or less at 1.2 times the set current and will not operate within 2 h at 1.05 times the set current.
- A device for direct temperature control by embedded temperature sensor.
- Another equivalent device.

8.4.3 Emergency Switch Off

In an emergency, there must be multiple means for shutting down the electricity in a hazardous area from a safe location. Separate circuits must be provided for electrical apparatus in order to prevent additional dangers.

8.4.4 Electrical Isolation

All phase and neutral conducts must be isolated by the electrical isolators. Maintenance activities requiring power isolation should not be performed without a proper work permit. The process of restoring supply must be undertaken with great care. Ex equipment must be properly maintained in order to ensure that it is ready for service, as loose bolts of enclosures, unplugged entries, inappropriate fittings, etc., render Ex equipment unsuitable for service.

8.4.5 *Overhead Lines*

It is not recommended to use overhead lines in hazardous areas. Wiring in conduits or cables should be used for all power distribution.

8.4.6 *Cable and Conduit System*

- Cables and conduits must be connected to electrical apparatus in accordance with the relevant protection requirements.
- The cable and conduit system shall be suitable for environmental conditions, such as chemical, vibration, heat, and humidity.
- To prevent the passage of flammable media when crossing over from a covered area to an open area, trunking, ducts, pipes, trays and pipes carrying cable and conduits shall have no prorogating barriers. Area classification shall not be vitiated by such cross-overs.
- Aluminum conductors of less than 16 mm^2 and without lugs shall not be used in other than intrinsically safe circuits.
- Non-sheathed single-core cables shall be used within enclosures and conduits. These are only appropriate for use in wiring intrinsically safe system.
- Cables employing thermoplastic materials having cold flow characteristics shall be used with suitable cable glands. Cable glands with compression seals are not suitable for such cables.
- Physical contact between the armor/metal sheath of a cable and pipe work or process equipment containing flammable media shall be avoided except in the case of heat trace.
- Cables shall be terminated using cable glands providing adequate clamping force. Optionally additional clamping elsewhere shall be provided unless not practical. In no case cables shall be used without cable clamping elsewhere if cable is being terminated using cable glands providing no clamping force.
- As far as possible, cable runs should be joint-less in hazardous areas. Cable joints if required shall be made through an enclosure of appropriate Ex protection. If the jointing is not subject to mechanical stress, epoxy-filled jointing kits can also be used.
- Surface temperatures of cables should not exceed the temperature class for the installation.
- Cable surface temperature shall not exceed the temperature class for the installation.

- Conductor connections shall be made using lugs, secured screw connectors, soldering, welding, or brazing with the exception of those in flameproof conduits or intrinsically safe circuits. Multi-stranded conductors, however, should not be terminated without lugs.
- It is recommended to terminate unused cable cores into spare terminals or connect them to ground. Ends should not be left unterminated or insulated with tape.
- Conduit and cable entries that are not in use in the apparatus should be plugged in.

8.4.7 Types of Cables

It is permissible to use mineral-insulated metal/thermoplastic/thermosetting/elastomeric or tough rubber/polychloroprene sheathed cables for fixed apparatus.

For portable apparatus, heavy rubber/polychloroprene/elastomeric sheathed flexible cables are permitted. As required, these cables may have flexible metallic armor or screen and protective earthing conductors (PE). These cables should be of adequate conductor size but not less than 1.0 mm^2. All cables shall have flame propagation properties as per IEC 60332-1:2004 [17].

8.4.8 Conduit System

The enclosures should be equipped with sealing conduits/stopping boxes adjacent to the enclosures so as to maintain an appropriate degree of ingress protection. Filling compounds used for sealing should not shrink after setting.

The conduit system used as a protective conductor must have a threaded junction capable of carrying fault current.

The conduit must be protected from both ambient and galvanic corrosion.

Furthermore, conduits associated with flameproof enclosures must be screwed with heavy gage seamless or seam-welded steel. The use of ordinary conduits is not permitted. Metal conduits that are flexible may be permitted.

8.4.9 Additional Requirement for Type of Protection Ex 'd'

In order to ensure the safety of Ex 'd' equipment with a flange joint, the installation must be carried out in a manner that ensures that there are no solid structural obstacles located near the flange joint (Fig. 8.2).

Flameproof joints must be protected from corrosion and water infiltration. Gaskets may only be used when permitted by the equipment documentation. The use of

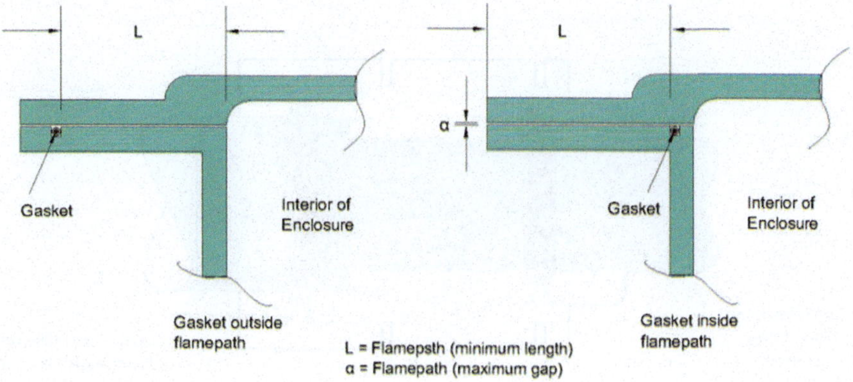

Fig. 8.2 Flange joint

silicon grease is recommended for preventing corrosion at flange joints. The flange joint should not be *tapped* for preventing the ingress of water.

8.4.10 Cable Glands for Ex 'd' Equipments

The electrical wires/cables connected to the apparatus are fitted using double compression flameproof cable glands. A cable gland may be tested and certified flameproof separately. It should sustain cable pullout test and torque test. The impact test is also required for the cable gland after fitting to the enclosure. A cable gland comprises a threaded nipple for attachment with flameproof enclosure, body consisting sealing (neoprene) ring for compression between cable sheath and itself, a cone and ring for clamping armor wires of the cable, and an outer cap with a rubber bush for weatherproof protection against ingress of water and dust into enclosure (Fig. 8.3).

8.4.11 Additional Requirements for Motors

It is required that all motors that use variable voltage or frequency or a combination thereof be type tested with drives. This is in order to ensure that no temperature rise violates the motor's temperature classification.

- It is recommended that motors be equipped with overload protection devices in order to ensure that they trip within a specified period of time.
- Motors should be equipped with single-phase protection to prevent temperature increases caused by phase loss.

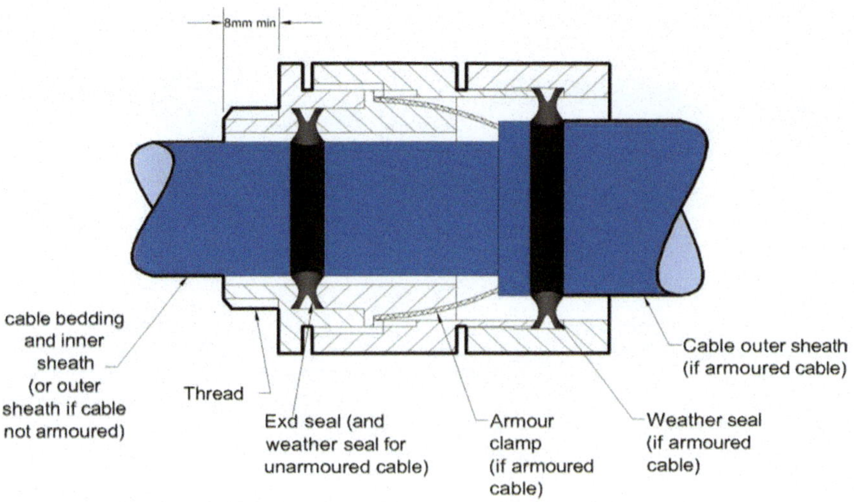

Fig. 8.3 Flameproof double compression cable gland

- The temperature of critical and high-capacity motors should be monitored on a regular basis.
- A motor's temperature classification must be maintained by adequate overload protection or temperature monitoring devices, or both.
- All terminals on Ex 'e'/Ex 'n'/Ex 'i' equipment must either be individually certified or certified along with the equipment.
- To ensure that the heater does not violate the temperature classification, it must be equipped with a temperature control device (Fig. 8.4).

8.4.12 Installation of Electrical Equipment in Gassy Coal Mines

Underground coal mines are governed by the Coal Mines Regulations, CMR 1957 (16), and the Central Electricity Authority (CEA) 2010 (17).

In accordance with Regulation 181 (3) of the Coal Mine Regulations of 1957, flameproof, enhanced safety, and intrinsically safe electrical apparatus and cable can only be used in underground coal mines if approved by the Chief Inspector of Mines, who is also designated by the Directorate General of Mines Safety.

The underground coalmines are classified into degree I, degree II, and degree III gassiness. According to the coal mine regulations 1957, the degree of gassiness depends on the volume of methane gases present per cubic meter of coal. Underground coal mines generally use flameproof and intrinsically safe products. These flameproof and intrinsically safe products are manufactured in accordance with relevant Indian codes. Any recognized test house, such as CSIR-CIMFR, Dhanbad,

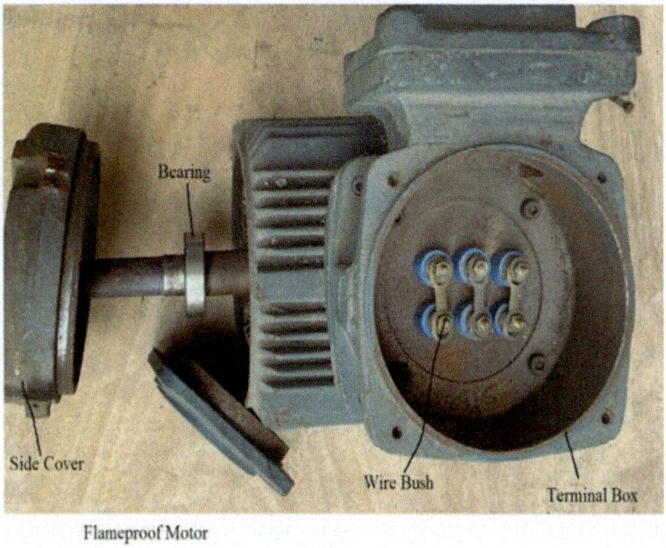

Fig. 8.4 Flameproof motor

ERTL, and Kolkata, must test these prototypes before they are used in underground coal mines. After receiving certification from the testing laboratory, the manufacturer submits the certificate to the Bureau of Indian Standards (BIS) for licensing purposes. Upon obtaining a license from the Bureau of Indian Standards (BIS), the manufacturer submits a test report as well as the certificate of the license granted by the BIS to the Directorate General of Mine Safety (DGMS). Once the equipment has been approved, it can be used safely in coal mines in India.

The Indian Electricity Rules require flameproof products to be installed within 270 m of the working face in gassy mines of Degree III. All signaling bells, telecommunications, and remote control circuits must be intrinsically safe. The installation of electrical products in degree II mines should be carried out at a distance of 90 m from the working surface. Flameproof and intrinsically safe products can be installed in any location within a gassy mine of degree I, which is in proximity to the last ventilation line.

In order to introduce continuous mining technology in Indian mines, several provisions of the Indian Electricity Rules have been amended. In situations where there is a lack of space, direct entry of cable can also be used for transportable machines. In addition, battery-operated shuttle cars may be used in degree I gassy mines for increased safety. It was necessary to make these amendments in order to introduce continuous mining technology to Indian mines in order to increase the production of coal there. Original rules call for flexible cable lengths of 90 and 180 m, but the amendments allow flexible cable lengths up to 250 m.

The proper selection of Ex equipment should be done to increase safety of installation. The installation and importance of Ex equipment are need to understand for better safety.

Statutory guideline is available to all the technical persons responsible to maintain safety of electrical installations. However, sound engineering knowledge and experience will ensure proper selection and installation of the equipment installed in hazardous areas.

References

1. Singh AK (2011) Safety requirements and selection of electrical equipment for CMM recovery. First Indo-US workshop on Coal Mine Methane, CSIR-CIMFR, Dhanbad, pp 129–134
2. Vishwakarma RK, Ahirwal B, Kumar A, Kumar N, Mandal HK, Singh AK (2014) Recent trend and future scenario of ex-equipment for use in gassy mines. In: 2nd international seminar DTEX-2014, pp 69–82
3. IEC 60079-1:2014 (2014) Explosive atmospheres-part 1: equipment protection by flameproof enclosure 'd'
4. IEC 60079-0:2017 (2017) Electrical apparatus for explosive gas atmospheres part 0: general requirements
5. IEC 60079-11:2011 (2011) Explosive atmospheres, part 11: equipment protection by intrinsic safety 'i'
6. I EC 60079-2:2014 (2014) Electrical apparatus for explosive gas atmosphere part 2: pressurized enclosure 'p'
7. IEC 60079-5:2007 (2007) Electrical apparatus for explosive gas atmospheres part 5: powder filling 'q'
8. IEC 60079-6:2015 (2015) Electrical apparatus for explosive gas atmospheres part 6: oil immersion 'o'
9. IEC 60079-18:2014 (2014) Electrical apparatus for explosive gas atmospheres part 18: encapsulation 'm'
10. IEC 60079-15:2017 (2017) Electrical apparatus for explosive gas atmospheres part 15: increased safety/non-sparking 'e/n'
11. IEC 60079-7. Electrical equipment for explosive gas atmospheres—increased safety "e"
12. IEC 60079-20-1:2010 (2010) Classification of gases and vapours
13. Gupta BK (2014) Selection and installation of ex equipments. In: 2nd international seminar on design, development, testing and certification of ex equipment. DTEX 2014, pp 275–285
14. IEC 60079-17:2007 (2007) Explosive atmosphere (other than mines and explosives) part: 17 electrical installation, inspection and maintenance
15. IEC 60079-19:2006 (2006) Explosive atmospheres part 19, equipment repairing overhaul and reclamation
16. Pandey UN, Behera BN (2014) Electrical installation in hazardous areas. DTEX 2014, pp 3–10
17. IEC 60332-1:2004 (2004) Test of the fire behavior on an single core or a single cable (flame retardency)

Chapter 9
Inspection and Maintenance of Explosion-Proof Equipment

9.1 Introduction

The use of electrical equipment in a hazardous area is an important aspect. No matter how much effort is put into the classification of the area, the selection of equipment, or the installation of the equipment, a lack of maintenance can invalidate all that efforts in a matter of seconds. A missing bolt, for example, may compromise the protection of a flameproof enclosure and put the entire site at risk. It is therefore essential to pay close attention to the maintenance of electrical equipment in hazardous areas. It is recommended that users keep a detailed record of the location and condition of all electrical equipment. The record should include a log of any problems that occur, as well as details of any maintenance or repair work that has been done.

Hazardous areas require electrical installations (including electric components and systems) with special features and properties, to ensure that the safety aspect and safety requirements are met. The integrity of these special features must be maintained throughout the life of such installations. Periodic inspections and continuous supervision by skilled personnel are crucial to achieving this requirement. A maintenance plan is ultimately formulated based on these activities.

Inspection and maintenance of explosion-protected products should be performed by a trained individual who is familiar with the types of protection, the classification of hazardous areas, as well as the rules and regulations pertaining to hazardous areas. In accordance with the code of practice, personnel should be trained on a regular basis. It is the responsibility of end users to take care of these aspects in order to ensure the safety of the plant. This chapter systematically and elaborately describes the inspection and maintenance aspects of explosion-proof equipment (Ex equipment) as specified by IEC 60079-17 [1].

© The Author(s), under exclusive license to Springer Nature Singapore Pte Ltd. 2023
A. Kumar Singh, *Explosion-Proof Equipment in Hazardous Area*,
https://doi.org/10.1007/978-981-99-2516-2_9

9.2 General Requirements

9.2.1 Documentation

It plays an instrumental role in the inspection and maintenance of electrical equipment in hazardous locations. To accomplish the above task, the following documents must be kept up-to-date:

1. Hazardous Area Classification.
2. Temperature class and apparatus group.
3. A list and location of the apparatus, spare parts, technical information, and instructions provided by the manufacturer.

9.2.2 Qualification of Personnel

Inspection and maintenance of installations must be conducted by trained personnel who have received training in various types of protection, area classification, installation practices, and rules. Personnel must undergo continuing education or training on a regular basis. Training and experience claims must be backed up with evidence.

9.3 Type of Inspection

According to IEC 60079-17 [1], inspection types can be classified as follows:

- Regular periodic inspection.
- Continuous supervision by competent personnel.

The term 'continuous supervision' refers to plant maintenance activities that are performed by competent individuals on a regular basis. The periodic inspection interval is determined by the expected rate of deterioration of Ex equipment in hazardous areas (Fig. 9.1).

9.3.1 Initial Inspection

An initial inspection is required for all new installations before they can be put into service [2]. The purpose of this initial inspection is to ensure that the type of electrical equipment selected and its installation are suitable. In addition, it is needed when electrical equipment is replaced, repaired, modified, or installed. Changes in area classification will also require this procedure. It is not necessary to conduct an initial inspection of such parameters that are unlikely to be affected by the installation and

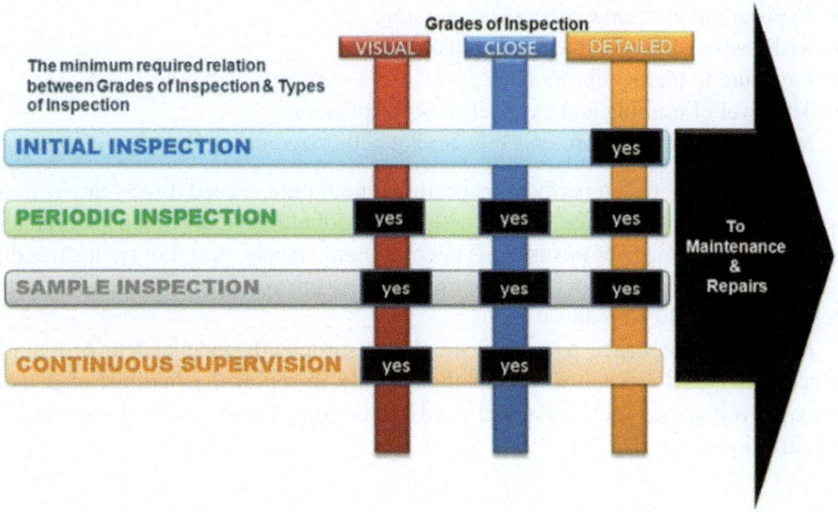

Fig. 9.1 Types and grades of inspection

have been inspected by the manufacturer. The flamepath of a flameproof motor, for example, does not require detailed initial inspection. However, the flamepath of the associated terminal box requires initial inspection after cabling and before fixing the cover.

9.3.2 Periodic Inspection

Periodic inspection is a routine inspection of all equipment. Inspections may be visual or close, leading to further detailed inspections if necessary.

The following factors are taken into account when determining the grade of inspection and the appropriate interval between inspections:

- The type of Ex equipment.
- Instructions provided by the manufacturer.
- Factors influencing the deterioration of Ex equipment.
- The zone of use.
- Inspection results from previous inspections.

The following factors determine the rate of deterioration of Ex equipment:

- The corrosion of metals.
- Chemical and solvent exposure.
- Dust and dirt accumulation.
- Ingress of water.

- Exposure to excessive process temperature.
- Risks associated with mechanical damage.
- Exposure to undue vibration.
- The level of training and experience of the personnel.
- Modifications or adjustments that are not authorized or improperly maintained.

The interval between periodic inspections should not exceed three years without seeking professional advice. For portable and handheld devices, which are typically susceptible to damage or misuse, this interval should be one year. For enclosures that require frequent opening or are prone to damage or misuse, such as battery housings, the interval should be reduced.

Periodic maintenance personnel should be sufficiently independent of maintenance activities so as not to impair their ability to report the findings of inspections. Nevertheless, such personnel need not be drawn from external independent organizations.

9.3.3 Sample Inspection

The purpose of sample inspections is to examine a portion of the installed electrical equipment. Sample inspections assist in determining the intervals and grades of periodic inspections. Their purpose is not to detect random faults, but rather to monitor the effects of environmental conditions, vibrations, design weaknesses, etc.

There are three grades of inspection: visual (V), close (C), and detailed (D). It is possible to perform visual and close inspections while the equipment is energized. To perform a detailed inspection, the equipment must be isolated from power and operation.

9.4 Grades of Inspection

The grades of inspection for the three types of inspection (initial, periodic, and sample) are as follows:

9.4.1 Visual Inspection (V)

An inspection that identifies defects, without the use of access equipment or tools, which are directly visible to the naked eye, such as missing bolts or broken glass. Below are some examples of visual inspections. An Ex ‘d’ flameproof equipment thread engagement is not properly engaged, and the cable entry is not plugged with an approved and certified flameproof plug. The cable should pass from the cable entry of flameproof equipment without using any flameproof cable glands.

Following points are to be considered during a visual inspection [3]:

1. All electrical equipment installed in hazardous areas should be certified by a testing laboratory.
2. The electrical equipment shall be installed/connected according to the certificate issued by the test laboratory.
3. In flameproof equipment, toughened glass should be used.
4. Code-compliant earthing.
5. Dust accumulation on enclosure surfaces should be avoided since this can affect the equipment's temperature classification.
6. Maintain a degree of dust and liquid protection for electrical equipment.
7. In hazardous areas, certified cable glands should be used.
8. There is a restriction on dust and water ingress.
9. A clearly visible warning sign.

9.4.2 Close Inspection (C)

Unlike a visual inspection, a close inspection incorporates all the aspects covered by a visual inspection, as well as identifying defects that are only apparent with the use of access equipment, such as steps (if necessary) and tools. It is not necessary to open the enclosure or de-energize the equipment for close inspections.

9.4.3 Detailed Inspection (D)

It can detect faults like loose terminations which cannot be detected by the close inspection but can be detected by the detailed inspection by opening up the enclosure or by using tools. When the inspection is completed in accordance with the planned/desired schedule, the risk of fire can be minimized while the life of the Ex equipment can be prolonged due to the high safety factor of the installation.

9.5 Important Points During Inspection

It is highly beneficial to use an independent inspector when inspecting Ex equipment since it is possible that the electrician(s) performing the maintenance may not be aware of the non-compliance. Below is a list of typical equipment and installation faults that may be noted during the inspection audit of Ex equipment for hazardous areas.

- The zone classifications have not been completed or are inaccurate as a result of a change in the process.
- Using non-explosion-proof equipment in hazardous environments.
- Despite certification for use in hazardous areas, the glands/adaptors are not explosion-proof or lack stoppers or blanking plates.
- Modifications made to the equipment without authorization—for example, additional cable entries drilled into flameproof (Ex 'd') junction boxes or additional holes drilled in increased safety (Ex 'e') fluorescent lights to facilitate mounting.
- There is no Ex certification on the equipment, and the conformity assessment has not been accepted.
- Ingress protection (IP) washers and gaskets have failed and no longer provide protection for Ex 'e' equipment.
- Cables and equipment are damaged.
- Gaskets (where required) are missing from flameproof equipment (Ex 'd').
- There are missing/loose bolts on flameproof equipment (Ex 'd').
- A faded and illegible Ex certification identity plate has been found.
- Although the equipment is certified for explosion protection for dust (DIP), it is not suitable for gas applications.
- Explosion-proof light fittings with cracked glass or missing seals.

9.6 Inspection in Operating Unit

When electrical equipment fails in a hazardous area, the following hazards are created:

1. There is a possibility of fatal accidents occurring as a result of electric shock.
2. Electric arcs, sparks, and hot surfaces may cause fire accidents and burn injuries.
3. An explosion caused by an electrical arc, a spark, and a hot surface in flammable media.
4. Loss of production due to an abrupt interruption in the operation of the plant.

9.7 Inspection of Various Types of Protected Equipment

9.7.1 Flameproof Ex 'd'

- Electrical equipment should be certified according to its area classification, gas group, temperature class, and environmental conditions.
- Electrical circuits are housed in flameproof enclosures that have been tested and certified.

- Flameproof enclosures must use toughened glass and be sealed with high-quality cement.
- Indirect and direct cable entry, bolts, and blanking elements are of the correct type.
- Flamepaths and gaps in a flameproof enclosure should meet code requirements. The faces of the flanges must be cleaned and the gasket should not be damaged.
- Lamp rating as certified by the test house.
- Motor fan should have sufficient clearance to enclosures and covers.
- Cables should not be damaged. Appropriate cables should be used.
- Epoxy should be used to fill the stopper box and cable box.
- The earthing, equipotential bonding, and insulation resistance should be satisfactory.
- Electrical equipment should be protected against corrosion and other adverse effects.
- The surfaces of the equipment should be free of dust and dirt.

9.7.2 Intrinsically Safe Ex 'i'

The following points should be considered during the inspection of Ex 'i' products:

- Installation of the equipment is done based on the certification granted by the testing laboratory, including area classification, gas group, temperature class, and environmental conditions.
- The circuits and ratings of intrinsically safe products cannot be modified without authorization.
- Barrier units, relays, and other energy-limiting devices are used in hazardous areas.
- There is a proper separation between intrinsically safe and non-intrinsically safe circuits.
- Cabling is installed in accordance with documentation, and cable screens are also earthed accordingly.
- Electrical connections should be tight on intrinsically safe products, and earth continuity should be satisfactory.
- Dust and dirt accumulation on equipment surfaces is not permitted since it affects the product's temperature classification.
- Ensure that the equipment is protected from corrosion, moisture, vibration, and excessive temperatures.

9.7.3 Increased Safety Ex 'e'

The following points should be considered during the inspection of Ex 'e' products:

- Increased safety products installed in hazardous areas must be certified by the testing laboratory for extent of hazard, gas group, temperature class, and environmental conditions.
- Unauthorized modification of products is not allowed.
- The accumulation of dust and dirt is prohibited on the surface of the equipment.
- The bolts, glands, drain plug, and entry plugs should be completely tight.
- The condition of the enclosure gasket should be satisfactory.
- Guards should be correctly fitted.
- Electrical equipment connections are tight and clearance should be maintained.
- Ensure that electrical equipment is completely protected against corrosion, moisture, vibration, excessive temperatures, and other adverse conditions.
- The temperature rating and type should comply with the testing laboratories certificate.

9.8 Maintenance

Maintenance actions are taken to prolong the life of equipment, machines, or systems by retaining or restoring them to their original operable state. Maintenance includes both corrective and preventive actions. Maintenance of Ex equipment must be performed in order to ensure that the products are connected properly at all times. IEC 60079-19 [4] governs the repair and overhaul of electrical apparatus in hazardous areas.

Following points should be taken care while maintenance of Ex equipments:

9.8.1 Basic Requirement of Maintenance

Following the inspection, the condition of the Ex equipment and installation should be noted, and any necessary measures should be taken. Maintaining the integrity of the type of protection for Ex equipment and installations may require consulting the manufacturer's instructions. Safety documentation must be followed when replacing parts. Ex equipment or installation should not be altered without authorization. Any changes that pose a threat to safety should not be made. Examples of such precautions include [5].

9.8.2 *Maintenance of Flameproof Equipment Ex 'd'*

These are the maintenance points of following flameproof products:

9.8.2.1 Flameproof Lighting Fitting

Maintenance of light fitting like well glass, flood light, tube light luminaries, etc. is used in surface industry like drilling rig, refineries, fertilizers, and underground coal mines. Round or dome-shaped glass is used for floodlight or well glass lighting fittings. Figures 9.2a–c show the different types of lighting fittings used in underground coalmines and surface industries like petrochemical, fertilizer, etc. The glass part is to be fixed with cementing compound like epoxy resin from inside only and further supported by the glass retaining ring or plate. The glass cement should not loose its properties in the entire range of service temperature. The glass part should be heat and impact resistant and of toughened quality. It may be provided with wire guard of maximum mess size 50 × 50 sq. mm. The minimum impact energy for use in coalmines is 7 J for glass parts provided without guard.

The following important points should be taken care during maintenance and installation:

(i) Check the hair crack on the surface of glass light fitting.
(ii) It is checked that fittings are certified with wire guard.
(iii) Cover of fitting should be completely tight.
(iv) In case of threaded joint, minimum five threads engage.
(v) Replace the same size lamp rating and type, it should not be changed, otherwise temperature class will differ.
(vi) Gasket should be replaced, and if it is damaged, it can impair the weatherproofness of the light fixture.
(vii) If gasket is not provided in the supplied product and even not mentioned in the certificates/report, it should not be provided by user without permission of testing authority because flameproofness will be affected.
(viii) The nipple or bushing connected between two enclosures to pass wires should not be removed and epoxy sealing should also be checked.

9.8.2.2 Maintenance of Flameproof Control Panel/Junction Box

When an empty flameproof enclosures are assembled with different types of electrical and electronic component to make a system with other associated apparatus, it may be flameproof and intrinsically safe or any other combination of protection. The combination of different flameproof enclosure is called flameproof control panel. The glass inspection windows should have sufficient cemented path and toughened and clear glass. All cables must be passed through flameproof-cum-weatherproof double compression cable gland of appropriate gas group. During the maintenance

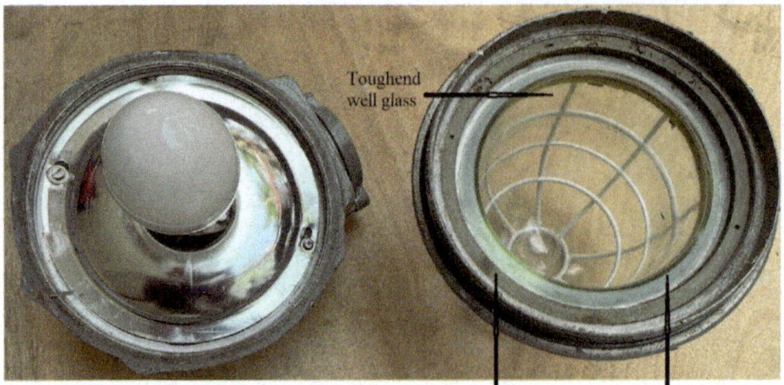

(a) Flameproof Well glass Fitting

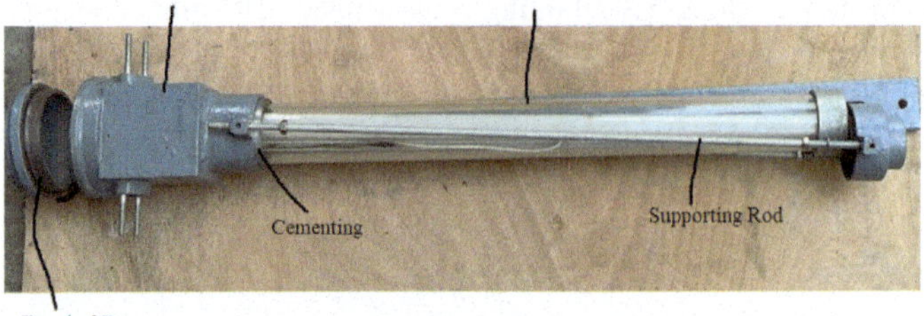

(b) Flameproof Tube light fitting

(c) Flameproof Flood light fitting

Fig. 9.2 **a–c** Different types of flameproof lighting fitting

of explosion-proof junction box, check the proper fastening and tightening of high-tensile strength (HTS) bolts. If such bolt is missing, it should be replaced with identical size. Normal/ordinary glass must not be replaced during repair. Enclosure should not be drilled additional for cable entry during maintenance. Figures 9.3a and b show the flameproof junction box and flameproof control panel.

9.8.2.3 Maintenance of Flameproof Cable Gland

Figures 9.4 a and b show the different components of flameproof double compression cable gland. Following points should be taken care during maintenance of cable glands:

- If the sealing ring of the cable gland is damaged, it shall be replaced with an identical sealing ring provided by the manufacturer.
- The size of cable should be inserted into the same size of cable gland.
- For explosion-proof products, only cable glands that have been certified and approved by a test laboratory should be used.
- Minimum five number of threads shall be engaged in nipple.

9.8.2.4 Maintenance of Flameproof Plug and Socket

Following points should be taken care during maintenance of plug and socket (Fig. 9.5).

- Check that plug and sockets are interlocked mechanically or electrically.
- They cannot be separated while the contacts are energized.
- The contacts cannot be energized when the plug and sockets are separated. Plugs with components remaining energized with a socket are not permitted.

9.8.2.5 Maintenance of Flameproof Stopping Plug/Closing Device

Ensure that the unused apertures/cable entries of flameproof enclosures are closed with a flameproof closing device. Check the fitment of the flameproof stopping plug, which can be fitted or removed from either the outside or inside of the flameproof enclosure. It is always necessary to maintain the flamepath (number of threads engaged) between the stopping plug and the enclosure wall (Fig. 9.6).

9.8.2.6 Flameproof Motor Ex 'd'

In hazardous areas, used motors must be inspected according to a certificate issued by a test laboratory. The test report specifies the flamepath, gaps, and gross volume of the motor. Figure 9.7 shows the different parts of flameproof motors. The flamepath

(a)

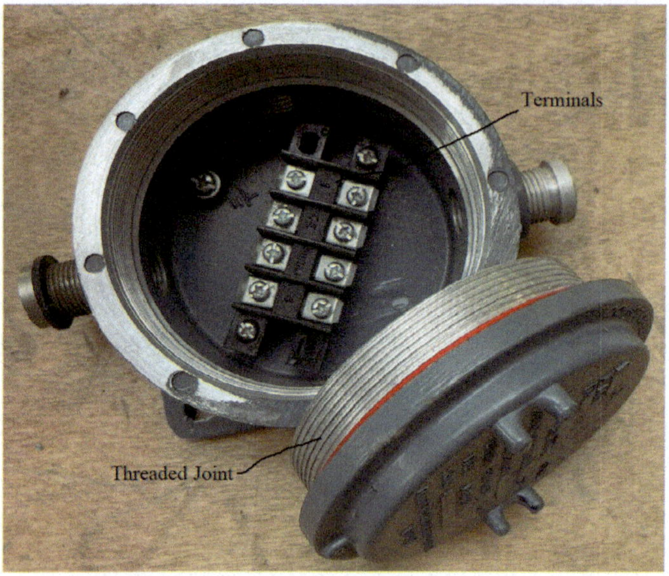

Flameproof Junction Box

(b)

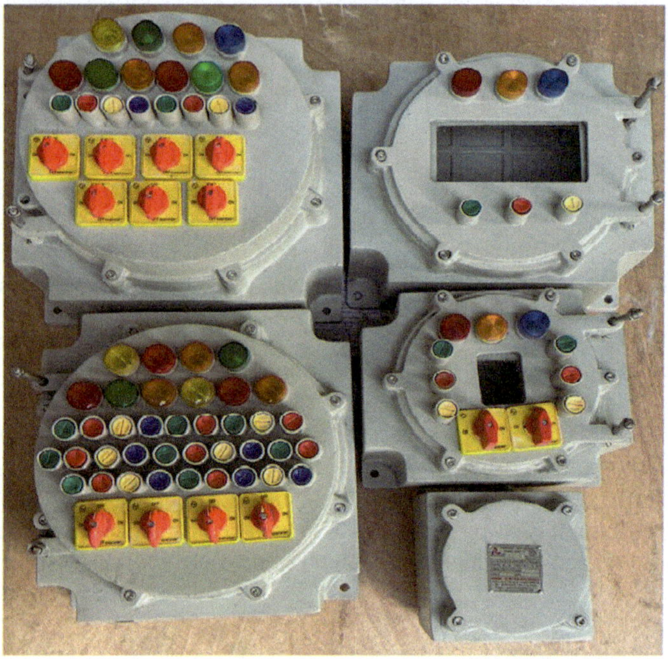

Flameproof Control Panel

Fig. 9.3 **a**, **b** Shows the flameproof junction box/control panel

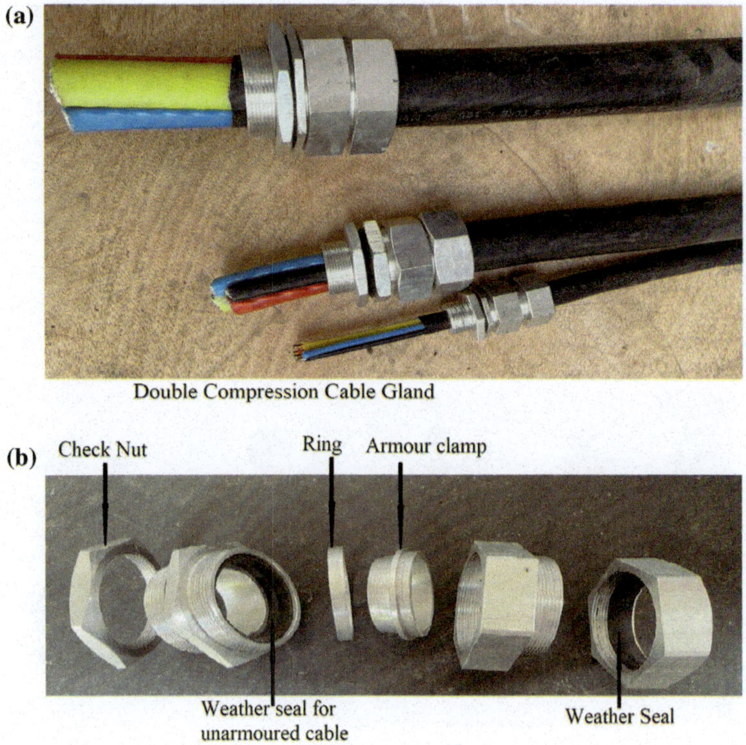

Fig. 9.4 **a**, **b** Double compression cable gland and its components

Fig. 9.5 Flameproof plug and socket

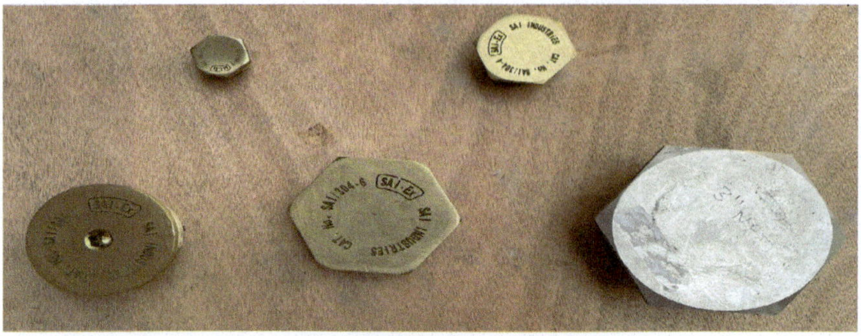

(a) Flameproof Stopping Plug

(b) Flameproof Reducer

Fig. 9.6 a, b Flameproof stopping plug/reducer

between rotating machine and gland is very important. Sleeve bearing joint is not applicable in IIC gas group for motor because it is not allowed due to non-radial clearance. The gap between the fan and its hood should be maintained.

9.8.3 Maintenance of Intrinsically Safe Circuits

An intrinsically safe circuit is one in which any spark or thermal effect produced under normal conditions cannot ignite the flammable medium. Using entity parameters such as limiting resistance, fuse, diode, capacitance, and inductance, this circuit prevents any explosions from occurring. During maintenance of intrinsically safe product, following points should be taken care of (Fig. 9.8):

- Input, output parameter of intrinsically safe circuit should be checked as per certificate issued by test laboratory.
- Temperature of total component of intrinsically safe product should be checked, and it should be below the temperature class granted by the laboratory.

Flameproof Motor

Fig. 9.7 Different parts of flameproof motors

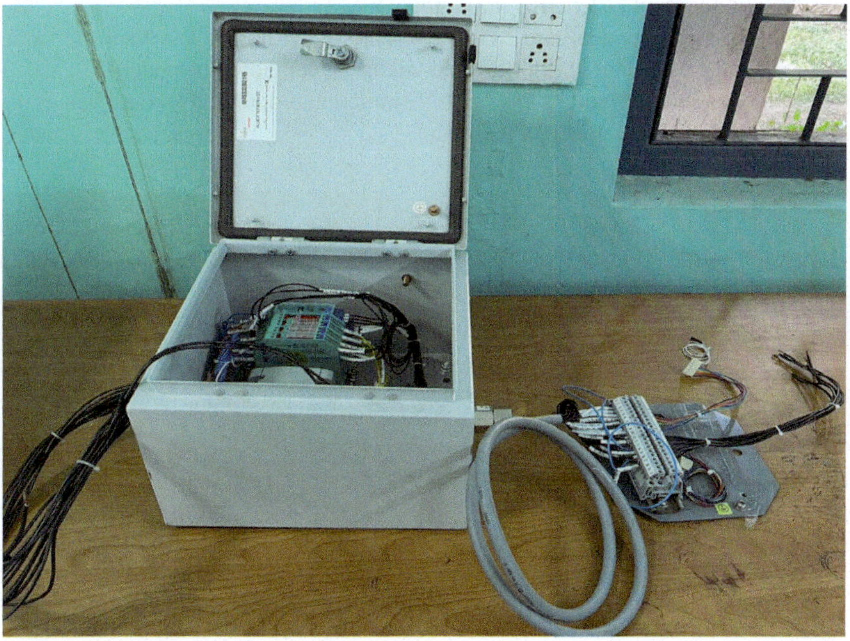

Maintenance of Intrinsic Safety Circuit

Fig. 9.8 Intrinsic safety circuit

- No accumulation of dust and water on the surface of the printed circuit board (PCB) of the circuit. It should be regularly monitored by the end user.
- Thermography by Thermal Imagine Camera of the intrinsically safe circuit to be find out the fault by heat analysis.

9.8.4 Maintenance of Increased Safety Ex 'e' and Non-sparking Ex 'n' Equipment

Explosion is prevented in hazardous area by eliminating arc, spark, and hot surfaces by increased safety and non-sparking product. Following points should be taken care during maintenance of these products (Fig. 9.9):

- Increased safety and non-sparking equipment protect against excessive temperature, arc, spark, water, dust, and impact.
- The cover should only be removed for a short period of time and equipment should not be left unsupervised without being protected.
- If dust or moisture has been deposited, it should be removed, and then the cover should be closed.
- Terminals for external connections should be dimensioned for extended connections and securely fastened to the conductors.
- The clearance between bare conductive parts must not be less than the values specified in accordance with the rated voltage specified in the code.
- The temperature of motor parts must be limited in order not to exceed values that could damage the stability of the motor.

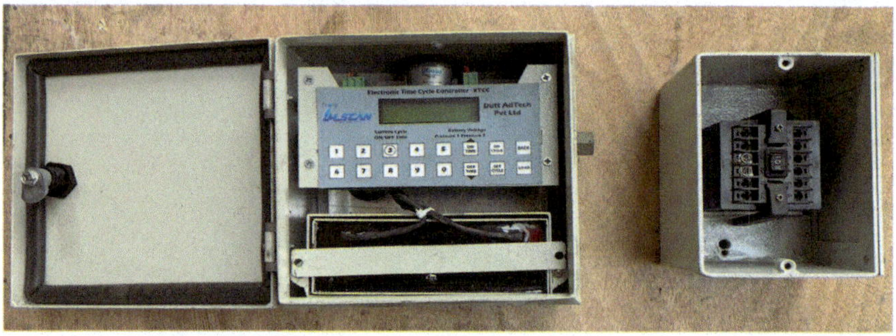

Increased Safety Junction Box / Control Panel

Fig. 9.9 Increased safety junction box/control panel

9.8.5 Maintenance of Pressurized Equipment Ex 'p'

This type of protected products prevents explosion in a hazardous area by keeping the flammable media out of the enclosure. During maintenance of these products, following points should be taken into consideration (Fig. 9.10):

- In the case of pressurized equipment, keeping a specified minimum overpressure within the enclosure is important because the safety of Ex equipment is dependent on it.
- The enclosure, ducts, and coupling components should be able to withstand an overpressure of 1.5 times the specified maximum overpressure or 2 mbar, whichever is greater for 2 min ± 10 s.
- During the maintenance, it should be remembered that Ex 'p' equipment having partial or fully purged system.
- The cover of the enclosure and the tubing should be of non-incentive material.
- The cables entered inside the pressurized enclosure should be well sealed with sealing compound.
- The inlet tubes for purging and the exhaust tubes for protective gas should be such that an overpressure of 5 mm WC remains maintained.
- Doors and covers of pressurized unit should be provided with special fasteners.

Fig. 9.10 Pressurized motor. **a** Purge control system. **b** Mechanical timer. **c** Vent system

- Check all cable entries and conduit connections sealing to maintain the ingress protection rating as well as to minimize the leakages of the protective gas.
- The maximum surface temperature of the apparatus should be less than the permissible temperature range.
- All audio alarm/annunciator should be checked properly.
- Pressure level gauge for low-/high-level protection should be checked properly.
- Check electrical and mechanical door interlocks in Ex p panel.

9.8.6 Important Points for Maintenance of Ex Equipment

During the maintenance of Ex equipment, the following points should be taken into consideration.

- Ex equipment should not be opened in the hazardous area in an energized state.
- A maintenance order should be issued by the competent authority of the plant or mine for Ex equipment in hazardous areas.
- During maintenance, unused holes or apertures should be plugged with certified flameproof plugs.
- Flameproof products should not be painted with aluminum paint.
- Any changes to tested and certified products are not allowed during maintenance in a hazardous area.
- The manufacturer should permanently mount an instruction plate near the apparatus.
- A safety audit by an Ex expert is very essential to notice minor issues that we might overlook.
- In hazardous areas, only certified and approved Ex equipment should be used.
- To control the surface temperature of the motor and its associated equipment, a calibrated, certified, and approved temperature sensor must be used.
- It is necessary to determine whether the lamp rating and type are correct and whether they are electrically protected.
- Make sure that all of the bolts, glands, and stopper boxes have been installed and tightened.
- There should be no leakage of the compound from the stopper or cable boxes.
- The motor/junction box should be checked to determine if excessive dust or dirt has accumulated.
- It is the user's responsibility to ensure that the motor is protected against overheating by providing overload and temperature detectors.
- The termination of supply cables to motor terminals must conform to the requirements of the relevant standards.
- Avoid using hammers and sharp tools, which can damage the flat joint surfaces of the enclosure.
- For opening cover bolts, always use the right size of Allen-key.

- Dry cells are used in the protected apparatus. It is important to ensure that replacement cells are of the correct type and voltage.

Inspection and maintenance of hazardous areas are an extremely important task. A qualified and experienced person is required to perform inspections and maintenance in hazardous areas. In order to increase the level of safety of installations in hazardous areas, the process should be properly organized and conducted in a timely manner.

9.9 List of IEC Code Related to Inspection and Maintenance of Ex Equipment

Personals who are involved in maintenance of Ex equipment must be familiar with the following standards:

Sl. No.	Standard	Title
1	IEC 60079-14:2014	Electrical apparatus for explosive gas atmospheres part 14: electrical installation in hazardous areas (other than mines)
2	IEC 60079-10:2021	Electrical apparatus for explosive gas atmospheres part 10: classification of hazardous area
3	IS 13408–1:2013	Code of selection installation and maintenance of electrical apparatus for use in explosive atmospheres
4	IS: 8239:1991	Classification maximum surface temperature of electrical equipment for use in explosive atmospheres
5	IEC 60079-7:2015	Electrical apparatus for explosive gas atmospheres part 7: increased safety 'e'
6	IEC 60079-0:2017	Electrical apparatus for explosive gas atmospheres part 0: general requirements
7	IEC 60079-11:2011	Electrical apparatus for explosive gas atmospheres part 11: intrinsic safety 'I'
8	IS: 8240:1976	Guide for electrical equipment for explosive atmospheres
9	IS: 2147:2014	Degrees of protection provided by enclosures for low-voltage switchgear and controlgear
10	IS: 4691:1985	Degrees of protection provided by enclosures for rotating electrical machinery
11	IS: 8241:1976	Methods of marking identifying electrical equipment for explosive atmospheres
12	IS: 8224:2003	Specification for electric lighting fitting for explosive atmospheres
13	IEC 60079-15:2017	Electrical apparatus for explosive gas atmospheres part 15: type of protection 'n'
14	IEC 60079-2:2014	Electrical apparatus for explosive gas atmospheres part 2: pressurized enclosure 'P'
15	IS: 2206-1:1984	Specification for flameproof electric light fixtures

References

1. IEC 60079-17:1996 Inspection and maintenance of electrical installations in hazardous area
2. Gupta BK (2009) Inspection and maintenance of Ex equipment. In: 1st international seminar and exhibition for explosive atmospheres on "Recent trends in design, development, testing and certification of ex-equipment". CSIR-CIMFR, Dhanbad (DTEX 2009) 29–31 October 2009
3. Kumar N (Inspection and Maintenance of Ex Equipments for Hazardous Area) (2017) Executive development programme on "FLP equipment" for ONGC executive. March 2017, CSIR-CIMFR, Dhanbad, pp 27–31
4. IEC 60079-19
5. Ahirwal B, Singh AK, Vishvakarma RK, Singh VK (2004) General maintenance requirements for hazardous area electrical equipment. Mine Mineral Rep 1(1):36 (April 2004)

Chapter 10
Frictional Ignition Hazard

10.1 Introduction

When coal cutting picks of road header, continuous miner, shearer, etc. strike with coal mass in underground coal mines, these picks do not ignite the firedamp and coal dust cloud, but when these strike with certain rocks, particularly sandstones, frictional heating may cause ignition of firedamp and coal dust cloud. The risk of ignition is greater if the cutting speed is high and machines that are made especially for cutting into rock are designed to have lower pick speeds. Figure 10.1 shows the continuous miner in Sarpi underground coal mines [1].

Most of the firedamp ignitions in coal mines occur as a result of coal cutting machines meeting. Petrological examination and rubber tests on colliery rocks have shown that rocks (other than pyrite) that give ignitions generally contain more than 30 percent by volume of quartz particles greater than 10 μm and that the quartz particles are usually larger than 70 μm. When the cutting speed is high, there is a significant risk of ignition. According to investigations conducted in other countries like South Africa, the UK, and MSW, this presents a significant risk during mining operations. This chapter attempts to explain the risk of explosion and fire caused by frictional ignition hazards using a worldwide literature review.

Sir Humphrey Davy [2] (flame-safety lamp) had conducted research into the inflammability of different gases in 1815, but in 1889, the French firedamp commission encouraged research into lighting gas by striking with a pick on fragments of hard stone or ironstone nodules to produce sparks.

There were three main types of frictional ignition:

- Pick on rock, as caused by a cutting tool during mining.
- Rock on rock, as occurs in roof falls.
- Metal on metal, as in mechanical equipment.

Although all three types of incident can produce sparking or heat of sufficient intensity to ignite methane, initiate a methane explosion, or even, under the right

© The Author(s), under exclusive license to Springer Nature Singapore Pte Ltd. 2023
A. Kumar Singh, *Explosion-Proof Equipment in Hazardous Area*,
https://doi.org/10.1007/978-981-99-2516-2_10

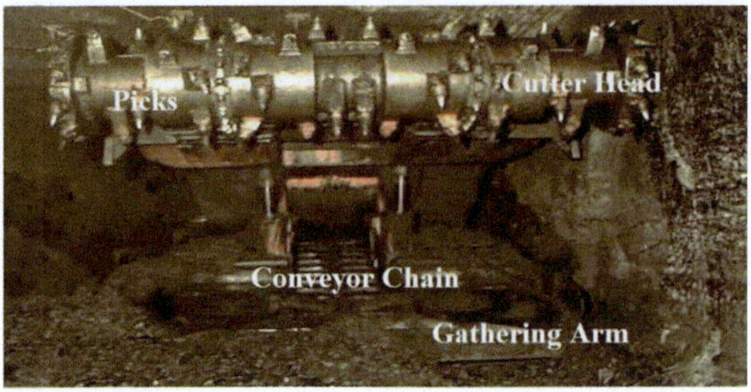

Fig. 10.1 Continuous miner in use at Sarpi underground coal mines

circumstances, lead to a devastating coal dust explosion, the analysis of South Africa, Phillips [3] clearly highlights the relative significance of each type of incident.

10.2 Known Facts Concerning Cutting Pick Ignition

10.2.1 Ignition Mechanisms

Researchers believed that sparks produced by picks could ignite the gas at an early stage. However, Blichensderfer et al. [4] and Burgess and Wheeler [5] have shown that sparks are not usually capable of igniting methane, which has a high ignition temperature (640 °C). In the 1940s and early 1950s, Wynn [6] proposed that ignition occurred via piezoelectricity. However, Powell et al. [7] demonstrated that ignition occurred well after the pick, and the delay was too long for an electrical discharge to be responsible. In various research reports, Powell has conclusively demonstrated that

- Ignition of methane air by a pick cutting rock occurs at or near the rock surface after the pick has passed.
- Ignition is caused by the hot material left behind by the pick and is likely to be at temperatures exceeding 1200 °C.

10.2.2 Rock Types Involved in Frictional Ignition

In a recent study by Ward et al. [8], it has been reported that high ignition risk can be associated with some conglomerates due to the presence of hard, homogenous, and often siliceous lithic fragments as well as, if not instead of quartz grains. Sparking has also been observed in rocks rich in feldspar, mica, and clay, but the phenomena of sparking and gas ignition are not necessarily related. The accepted facts are:

- The likelihood of ignition is greater for rocks containing more than 30% quartz, particle sizes greater than 10 μm (usually more than 70 μm), and that retain strength at temperatures at least 1250 °C.
- When pressed against a grindstone wheel, the rocks described above produce yellow-to-white heating, followed by the appearance of a molten quartz patch.
- When blunt or badly worn picks impact nodules of iron pyrite, the fine dust created oxidizes rapidly, and this self-heating can result in the ignition of methane. The strata must contain an appreciable percentage of iron pyrites before this occurs, and rock containing less than 10% iron pyrite can be considered safe, Blickensderfer et al. [4].

10.2.3 Effect of Cutting Speed

- There have been several studies that have examined the effect of the speed at which a pick moves through rock. This parameter has been investigated in relation to sharp, blunt, and pre-heated picks. Increasing pick speed results in a larger hot spot behind the pick, which increases the likelihood of ignition. In conclusion, we find that:
- A significant factor in determining the likelihood of ignition is the cutting speed. If the pick speed is reduced to 1.5 m/s or 1 m/s, the risk will be significantly reduced. Picks, both sharp and blunt, exhibit this phenomenon, but not pre-heated picks, where air cooling plays an important role.
- When cutter picks are used at low speeds, wear is minimized, and sharp picks are less likely to ignite methane–air mixtures.

Figure 10.2 shows the results of experiments carried out with single radial-type coal cutting picks. At selected cutting speeds, the cutting time to produce ignition of a 7% methane–air mixture was noted for several repeat experiments. Statistical analysis of the observations then provided a value for the cutting time to produce a 50% probability of ignition at each pick speed.

Most coal cutting machines operate at pick speeds between 3.0 and 4.0 m/s; Figure 10.2 shows that only a marginal increase in the cutting time for a 50% probability of ignition is obtained by lowering the pick speed within this range, but that a marked increase is achieved by lowering the pick speed to 1.25 m/s. An increase in ignition time indicates a safer condition.

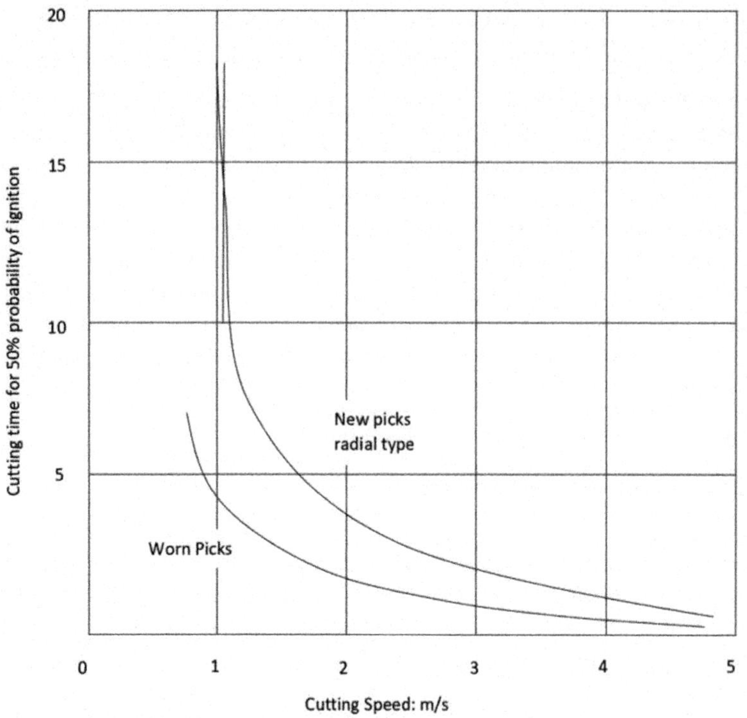

Fig. 10.2 Variation of cutting time for 50% probability of ignition with cutting speed

10.2.4 Effect of Water

Experiments on the use of water for the suppression of ignition have shown that a spray of water is more effective than jet, and that the most important factors governing the effectiveness of a spray are the region of application of the water, the droplet size of the spray, and the amount of water per unit area applied to the rock surface. Water applied above the pick, as for pick face flushing, has no effect on the probability of ignition: the most effective position for application is from behind the pick (Fig. 10.3).

The cutting time for a 50% probability of ignition for a single new pick cutting at 3.0 m/s, with no water, is 3.4 s. With a pick cutting at the same speed, but with a spray of water droplets of 100 μm surface mean diameter applied to the rock surface at a rate of 8.7 lit/m^2/s, the cutting time for a 50% probability of ignition is 15.8 s, which is the same as the value obtained with a new pick cutting at about 1 m/s with no water.

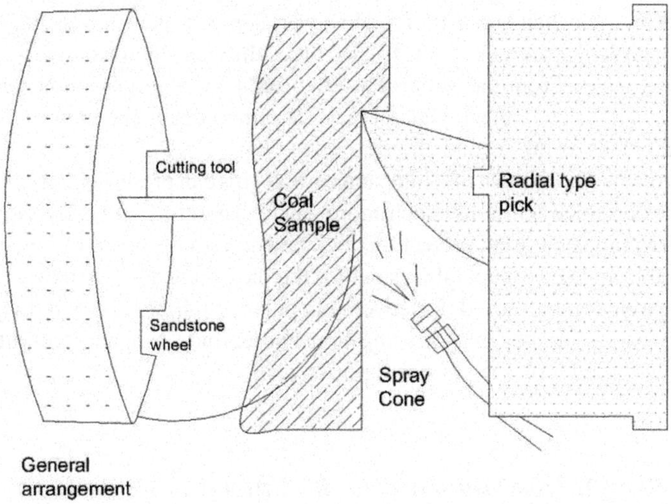

Fig. 10.3 Position of cutting picks

10.2.5 Cutter Pick Forces

- South African coal is harder than most of the country's coal. Therefore, the work rate per pick can range from 30 to 40 kW, and the cutting speed is approximately 3 m/s. These conditions generate high levels of stress between picks and the material being cut. A deeper cut with a higher power throughput may be more incendiary than shallower cuts with a lower power throughput, according to Lewis et al. [9], Larson et al. [10], and Roepke et al. [11].
- In order to make deeper cuts, a greater force is required, both in terms of friction and the normal force applied to the pick during cutting. As the normal force increases, the friction between the tool and the rock increases. This results in a hot spot forming on hard rocks, which creates the conditions necessary for pyrite to self-heat.
- By using sharp tools and optimizing their design, cutting forces can be significantly reduced. Therefore, tool geometry plays a significant role in reducing frictional ignitions.

10.2.6 Optimum Pick Design

In an attempt to reduce the frictional ignition hazard, Powell [12] has proposed design guidelines for coal cutting picks. Although the major coal-producing machine in the UK is the coal shearer and this is the type of machine at which the guidelines are aimed, some of the research findings should be just as applicable to continuous miners. According to Powell, 'limited evidence has shown that large radial attack picks appear safer than small tools'. However, some of the best results have been obtained with forward attack picks. The worst results were obtained with wide-bodied point attack picks. From the point of view of reducing the ignition hazard. Powell recommends a forward attack pick, at least 22 mm wide, as the preferred pick with wide radial picks as the second choice.

Experiments have shown that tungsten carbide tips can produce ignitions, but steel pick bodies produce ignitions much more easily. Therefore, it is advisable to have an oversized tungsten tip to protect the pick body. We have not yet found a material more suitable and economical than steel for the pick body. Powell also recommends tungsten carbide round-nosed tips covered in polycrystalline diamond. In her report, Courtney reports the benefits of using tungsten carbide-capped point attack tools (mushroom-tipped tools) to perform ignition tests.

10.2.7 Powell Has Summarized the Known Facts on Pick

- When a pick wears and runs on quartzite rock, it ignites. While ignitions can occur when the tungsten carbide tip rubs against the rock, most ignitions occur when the body or shank steel rubs against the rock.
- The mushroom-shaped tungsten carbide tips delay ignition, but cannot prevent it.
- When using tungsten carbide tips, the type or shape of the pick is important. Conical or pointed attack picks are inherently blunt and cause ignitions more easily than radial or forward attack picks with defined cutting edges.
- The wide-bodied point attack picks cause ignitions more readily than the narrow-bodied picks.
- A large or very large radial or forward attack pick is generally safer than a smaller one.
- Tips with rounded clearance faces are safer than those with angular surfaces.
- To cut rock safely, picks must stand up to rock much better than they do with existing tungsten carbide tips.
- The presence of a polycrystalline diamond layer on the rake face of a tip can greatly extend the life of a pick and decrease the likelihood of ignition even in the absence of water.

10.2.8 Rock on Rock (Ignition During Roof Falls)

The ignition of methane–air mixtures by roof falls is uncommon and, due to their location, is usually difficult to investigate. However, there have been numerous reports of sparks and bright flashes occurring during the collapse of a roof. In addition, there have been explosions that may only be reasonably explained by an ignition resulting from a fall or roof collapse. Several of these incidents have occurred in recent years since the introduction of roof bolting for strata control. Therefore, there are four possible sources of ignition.

- By breaking a bolt, a spark or hot surface is produced.
- Steel on steel impact, i.e., between broken bolt parts or between bolts and roof plates.
- Rock-on-rock impact.
- The impact of bolts on the surrounding strata (Figs. 10.4 and 10.5).

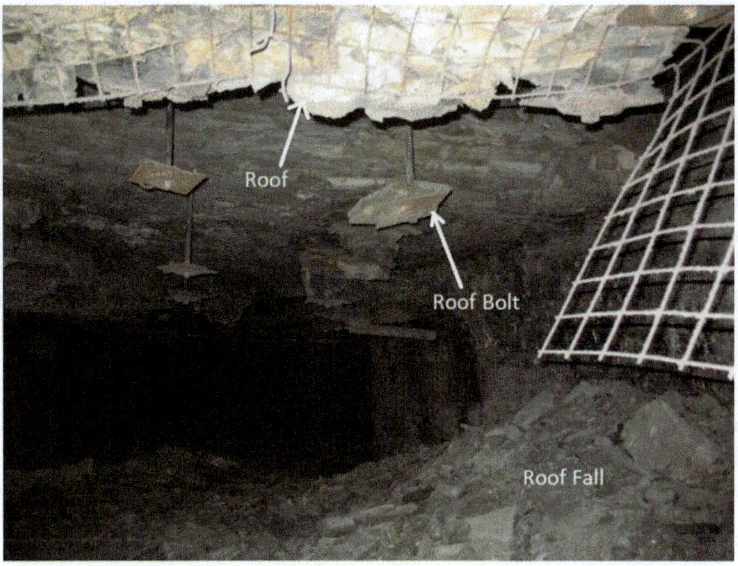

Fig. 10.4 Collapse of a roof due to roof bolt failure

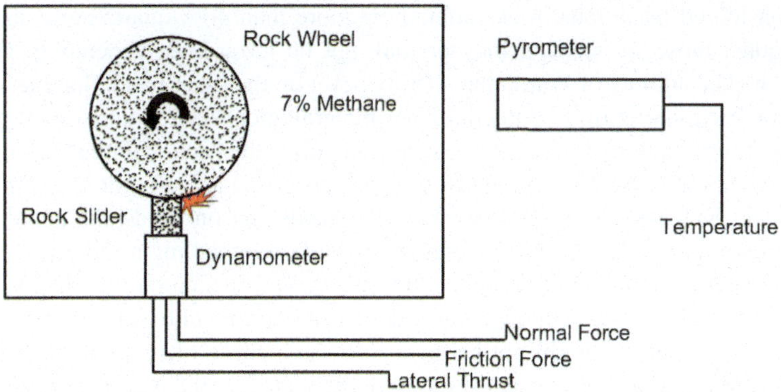

Fig. 10.5 Spark due to rock-on-rock impact

Titman has investigated the ignition of gas–air mixtures by firing steel balls at targets made of various materials in order to induce fracturing. Hydrogen–air mixtures could be ignited, but pentane–air and methane–air mixtures could not. According to other research regarding fractures in ocean-going oil tankers, 'it is generally accepted that the energy directly associated with the failure of the ship's steel would not be sufficient to cause the fracture surface to increase in temperature sufficiently to cause ignition'. Due to the high temperature necessary to ignite methane–air mixtures, Powell [13] has concluded that 'it is highly unlikely that any heating or sparks caused by fracture of a steel roof bolt will be sufficient to ignite methane-air mixtures'. The temperature of this fluid is highly dependent on the source of heat and its dwell time, but it is probably at least 640 °C.

If sufficient power is available, continuous rubbing of steel against steel can raise the temperature to at least the melting point of the metal. Therefore, it is not difficult to reach temperatures high enough to ignite methane. As described earlier, drill steel rubbing on a roof strap caused an ignition. Alternatively, the direct impact of steel on steel results in only a small temperature increase due to the deformation of the material, and any hot surface produced does not extend deep into the material, resulting in very rapid cooling. According to Powell [13], 'Impact of steel on steel during a roof collapse is also extremely unlikely to cause ignition'. Evidence suggests that for ignition to occur in drop-weight experiments, exceptionally hard steel (VNP > 700) must be involved.

It has been demonstrated that rocks with a high quartz content are capable of producing high temperatures when continuously rubbed. The research by Gray [14] has also shown that it is possible for a rock-on-rock impact to result in an ignition of methane in the air at velocities that can be encountered during a roof fall. Additionally, Ward [15] has recorded ignitions using a video camera and determined conclusively that the source of ignition is the hot spot that forms at the point of contact, rather than the sparks generated by the contact.

In the UK, rocks in the vicinity of all coal seams are regularly sampled and analyzed for quartz content. Generally, rocks containing less than 20% quartz are not considered hazardous; rocks containing more than 40% quartz are considered hazardous; those in between may or may not be hazardous, depending on their strength. The authors of Ward et al. [15] believe, on the other hand, that the assessment of frictional ignition potential from mineralogical data requires more than a simple determination of quartz content. To categorize the frictional ignition potential of rocks rubbing against each other, they have proposed a five-point scale. In 1960, Nagy and Kawenski [16] demonstrated that sandstone on sandstone could ignite methane in the air. However, fractures of a roof bolt or pulling a roof bolt through a washer and roof plate did not ignite the gas. As Powell summarizes the literature, 'Impact of rock on rock, although never demonstrated to ignite methane-air in drop-weight experiments, should not be discounted as a cause of ignition'. It has been noted that large slabs of rock can slide sufficiently far, i.e., 0.2 m, for ignition to take place.

During roof falls, steel, such as roof bolts and other forms of support or equipment, is the last point of contact. It has been known for centuries that this causes sparks that are incendiary. In a great deal of research involving cutter machine picks, rubbing experiments, and high-speed impacts of metal on rock, it has been repeatedly demonstrated that the contact of steel with rock can cause ignitions. Burgess and Wheeler [17], 1930 argue that a miner's hand pick can ignite methane–air when it strikes various sandstones. The energy generated by a man swinging a pick is sufficient to ignite methane in the event that blocks of sandstone fall from the roof onto a previously displaced roof bolt.

The most likely cause of an ignition caused by a roof fall is an impact between steel and rock.

10.2.9 Metal on Metal as in Mechanical Equipment

As a result of rubbing friction, such as that between a rotating steel wheel and a stationary piece of metal, or by impact, metal-to-metal ignition can occur (Fig. 10.6).

Based on research by Burgess and Wheeler [17] and others, it has been shown that in metal-to-metal contact, the more readily oxidizable metal determines the degree of ignition hazard. Hardness, melting point, ignition temperature, specific heat, heat conductivity, and brittleness of the metals are all factors that influence the size, duration, temperature, and heat capacity of incendiary sparks.

The friction between two iron or steel surfaces will only ignite methane–air mixtures if the friction is sufficient to raise the temperature of the steel surfaces to white heat. The sparks produced by friction between steel surfaces will not ignite methane unless they are concentrated by hitting an object. In coal mining, it should be remembered that friction in machine bearings and between steel surfaces is dangerous for a completely different reason. When coal dust is heated to about 1800 °C, it begins to smolder and can ignite.

Fig. 10.6 Spark due to metal-on-metal friction

During the 1940s and 1950s, a number of ignitions in European coal mines were attributed to the use of aluminum or magnesium alloys. A number of cases were observed where the rusted steel surfaces had been first smeared with aluminum before being subjected to rubbing friction under high bearing pressure. A hammer or a pick was believed to have struck the aluminum part in some cases.

As a result of this considerable body of research, the following conclusions have been reached:

- A material combination's incendivity is strongly influenced by the test conditions. The presence of more flammable atmospheres, higher velocity or energy of impact, and higher bearing pressure during frictional contact can all contribute to an increase in the Incendivity of a given pair of materials and may lead to ignitions where none were observed under less severe conditions.
- It is very easy for aluminum alloys to cause frictional ignition when they come into contact with rusty steel.
- Generally, harder aluminum alloys are more incendiary than softer alloys, and adding several percent of silicon to the alloy further exacerbates the problem. In similar tests, steels containing high silicon contents also caused ignitions, although with a much greater energy input than aluminum.
- When struck by a hard object, even aluminum paint on rusted steel can ignite.

Based on the findings of the research described above, several conclusions can be drawn that can contribute to improving coal mine safety. The following are among them:

- It is unlikely that friction between steel and steel will result in frictional ignitions of methane in the air. However, it should be avoided due to the potential for igniting coal dust and causing a fire.
- The use of high silicon steel should be avoided, as well as the prevention of rust on the surfaces of steel roof supports and other structures. Galvanized steel or zinc metal sprays or zinc chromate coatings can be used to accomplish the latter.
- Coal mines should not use aluminum paints on steel surfaces
- The use of light alloys, particularly magnesium-based alloys, should be restricted or prohibited. In the event that their limited use is permitted, all light metal alloys should be handled with care to avoid friction and impact with steel and hard rocks.

10.3 Brief Survey Report of Coal Mining of Four Countries

In coal mining, explosions caused by gas ignition are among the most feared hazards. In order for such an ignition to occur, two independent and random factors must be present simultaneously: the incendiary source and the explosive gas mixture. As a result of this simple concept, the probability of ignition is equal to the product of the probability that individual factors will occur. This concept illustrates that contributory risks are associative in nature and that the reduction of a single one has a direct impact on the total risk. No incendiary source means no risk, even if gas is present.

To determine where improvements in safety measures have reduced the risk and where new problems have arisen as a result of the deployment of new technology or changes in mining techniques, it is important to examine the changing pattern of ignition sources in our coal mining industry. The data for South African incidents over the past three decades have been analyzed and are presented in Table 10.1 along with the data for the first three years of the 1990s Phillips [18].

There are two very apparent changes: the steady increase in frictional ignitions following the introduction of continuous miners (CM) in the early 1970s and the decrease in the percentage of incidents caused by blasting. Throughout the period from 1960 to the present, electricity, lightning, naked flames, etc., have remained small and almost constant components of the overall scene.

A detailed analysis of the past decade provides a more realistic view of the current situation, with ignition sources grouped into three major categories, namely frictional, flame, and electrical. A friction ignition is caused by the cutting picks of continuous miners or road headers contacting stone, the cutting picks of coal cutters, and the cutting picks of shearers or ignitions caused by stone-on-stone or metal-on-metal contact. Among the causes of flames and ignition is blasting, spontaneous combustion, and heated surfaces. There were ignitions caused by electric sparks and lightning that formed stray currents due to electric sparks and lightning. According to Table 10.2, there have been a number of sources of ignition over the past decade and the frequency with which they have been involved.

It is evident that friction plays a dominant role as an ignition source, responsible for 32.5% of all incidents. Since many of the ignitions and explosions where the cause was never determined may have been the result of friction, it may be noteworthy to

Table 10.1 Changing pattern of ignition sources

Ignition source	Period							
	1960s		1970s		1980s		1990–1992	
	I	%	I	%	I	%	I	%
CM picks	0	0	1	5.5	14	29.5	6	25.5
CC picks	1	4.5	3	17	11	22.5	0	0
Shearer picks	0	0	0	0	1	2	0	0
Stone on stone	0	0	0	0	4	8	1	4
Blasting	14	64	6	33.50	5	10	1	4
Spon comb	0	0	0	5.5	0	0	2	8
Heated surface	1	4.5	1	5.5	1	2	0	0
Naked flame	2	9	1		0	0	0	0
Electricity	2	9	2	11	5	10	0	0
Lightning	2	9	4	22	3	6	0	0
Unknown	0	0	0	0	5	10	14	58.5
Total	22	100	18	100	49	100	24	100

I = number of incidents of ignitions/explosion

Table 10.2 Sources of ignition in South African collieries for the period 1984–1993

Source of ignition	Ignitions	Explosions	Total	% Total
Frictional				
CM picks	8	3	11	21
CC picks	2	0	2	4
Shearer picks	1	0	1	2
Stone on stone	3	0	3	5.5
Flame/heated surface				
Blasting	0	2	2	4
Spontaneous combustion	0	1	1	2
Heated surface	1	0	1	2
Electric				
Electric sparks	1	2	3	5.5
Lightning stray current	0	1	1	2
Unknown	**26**	**2**	**28**	**52**

note that frictional ignitions accounted for 68% of those incidents where the cause was determined. Alternatively, blasting and electric sparks account for only 8% and 12% of known ignition sources, respectively.

The frictional ignition problem is well documented in countries with a long history of mechanized mining. Despite the decline in the size of the industry, Browning [19] has documented a growing trend of such incidents in the UK (Table 10.3).

As Hardman [20] points out, the increase in frictional ignitions from the mid-50s to the mid-60s coincides with the increasing use of shearers. Although the number of frictional ignitions caused by cutting has decreased since the late 1960s,

Table 10.3 Ignitions in the UK 1951–1990

Different ignition sources						Frict. (%)	Frict./100 (Mt)
Years	Shot-firing	Electrical	Cutting frict.	Other	Total		
1951–54	10	8	8	29	55	14.5	0.9
1955–58	15	4	13	49	81	16.1	1.6
1959–62	25	12	35	23	95	36.8	4.6
1963–66	25	9	51	10	95	53.7	7.0
1967–70	14	9	71	6	100	71.0	11.7
1971–74	11	5	52	1	69	75.4	11.3
1975–78	6	1	60	6	73	82.2	13.7
1979–82	3	4	57	4	68	83.8	13.1
1983–86		3	33	7	43	76.7	10.3
1987–90		1	34	3	38	89.5	11.4

Table 10.4 Reportable ignitions of methane gas in NSW underground coal mines 1/7/87–31/12/93

Type of incident	Number
Continuous miner drum	12
Continuous miner shovel	1
Continuous miner cable	1
Longwall shearer drum picks	4
Rock fall in goaf	1
Cutting and welding operations	1
Roof drilling/bolting operations	2
Filling abandoned shaft	1
Total	23

the occurrence per 100 million tons mined has remained virtually unchanged since 1967.

During the period July 1, 1987–December 31, 1993, 23 ignitions of methane were investigated by the Department of Mineral Resources of New South Wales [21]. Based on the following table, it is possible to determine the location and type of incident (Table 10.4).

The data for NSW reveal that all types of frictional ignition account for 20 out of 23 ignitions. In addition, the friction generated during cutting action dominates the statistics with 16 out of 23 ignitions or 70% of the incidents. A total of 124.4 Mt was mined continuously during the period covered by this survey. As a result, the frictional ignition rate for longwalls is 3.22/100 Mt, the frictional ignition rate for continuous miners is 7.29/100 Mt, and the overall rate is 5.58/100 Mt.

According to a brief survey of underground coal mining in four countries, mechanized methods carry an inherent risk of sparking caused by machine picks. Despite nearly a century of research, incidents of this nature occur regularly throughout the world. As coal seams in South Africa are significantly harder than the average, cutting machines must be powerful, and drum rotation speeds must be relatively high in order to achieve acceptable production levels. There is a high likelihood of frictional ignitions occurring when a sandstone roof, floor, or inclusions in the seam are present; therefore, a study is warranted to identify the most appropriate preventative measures.

References

1. Samanta S et al (2021) Deployment of continuous miner in under ground coal mine—a case study of sarpi mine
2. Dav-1, Sir H (1815) Memoir on the ignition of mixtures of gas and air by electric sparks. Minutes of Royal Society of London, 9 Nov 1815

3. Phillips HR (1995) Coal mine explosion the South African experience. In: Proceedings of the 26th international conference of Safety in Mines Research Institutes, Katowice, Poland, vol 2, pp 89 102 (ref. no. 0098), Sept 1995
4. Blickensderfer R (1995) Methane ignition by frictional impact heating. Combust Flame 28:143–151
5. Burgrels MJ et al (1929) The ignition of firedamp by heat of impact of metal against rock. Safety in Mine Research Board, Sheffield Paper 54, London HMSO, p 25 (ref. no. 0070)
6. Wynn AHM (1952) The ignition of firedamp by friction. In: Safety in mines research. Rep. no. 42, Sheffield. SMRE, p 23 (ref. no. 0022), July 1952
7. Powell F et al (1975) The ignition of methane air by machine picks cutting into rock. In: Proceedings of XII international conference on coal mine safety research, Washington D.C., 1975, U.S. Bureau of Mines, OFR 83(1) 78, Sec. III, pp 7.1–7.10 (ref. no. 0079)
8. Ward CR et al (1991) Assessment of methane ignition potential by frictional processes from Australian coal mine rocks, 13. Mining Science and Technology, pp 183–206 (ref. no. 0086)
9. Lewis WT, Smith PR, Powell F (1983) Circumstances generating incendive sources in rock cutting. In: 20th international conference on mine safety research, Sheffield, U.K., I.D., p 10 (ref. no. 0080), Oct 1983
10. Larson DA, Dellorfano VW, Wingquist CF, Roepke WW (1983) Preliminary evaluation of bit impact ignition of methane using a drum type cutting head. United States of Bureau of Mines, R.I. 8755, p 23 (ref. no. 003)
11. Roepke WW, Hanson BD, Longfellow CE (1983) Drag bit cutting characteristics using sintered diamond inserts. United States Bureau of Mines, R.I. 8802 (ref. no. 0055)
12. Powell F (1991) Design guidelines for picks. Colliery Guardian, July 1991
13. Powell F (1994) Ignition of methane air during roof falls. Colliery Guardian 242(1):25–35
14. Gray I, Golledge P, Davis R, Sttory R. Frictional ignition caused by rock on rock impact in gas air mixtures, pp 35–43 (ref. no. 0093)
15. Ward CR, Cohen D, Panich D, Crouch A, Schaller S, Dutta P (1991) Assessment of methane ignition potential by frictional processes from Australian coal mine rocks. Min Sci Technol 13:183–206
16. Nagy J, Kawenski EM (1960) Frictional ignition of gas during a roof fall. United States Bureau of Mines. R. I. 5548, p 11 (ref. no. 0060)
17. Burgess MJ, Wheeler RV (1930) The ignition of firedamp by the heat of impact of hand picks against rock. Safety in Mines Research Board, Sheffield, paper no. 62, p 25 (ref. no. 0071)
18. Phillips HR (1995) Coal mine explosions of the South African experience. In: Proceedings of the 26th international conference of Safety in Mines Research Institutes, Katoowice, Poland, vol 2, pp 89–102 (ref. no. 0098), Sept 1995
19. Browning EJ (1988) Frictional ignition. In: Fourth international mine ventilation congress, Brisbane, Queensland, p 9, July 1988
20. Hardman DR (1993) Prevention or suppression of frictional ignitions. A discussion document. CSIR Miningtek. Report no. MT 7/93, p 12 (ref. no. 0085), May 1993
21. Department of Resources of New South Wales. Review of reportable frictional impact between aluminium alloys and rusted steel. United States Bureau of Mines, F.I. 8005, p 30 (ref. no. 0014)

The manufacturer's authorised representative in the EU is Springer Nature Customer Service Centre GmbH, Europaplatz 3, 69115 Heidelberg, Germany. If you have any concerns regarding our products, please contact ProductSafety@springernature.com

Printed and bound by CPI Group (UK) Ltd, Croydon, CR0 4YY
01/10/2026
02225900-0011